5th International Workshop on New Computational Methods for Inverse Problems (NCMIP 2015)

Journal of Physics: Conference Series Volume 657

Cachan, France
29 May 2015

ISBN: 978-1-5108-1731-9
ISSN: 1742-6588

Printed from e-media with permission by:

Curran Associates, Inc.
57 Morehouse Lane
Red Hook, NY 12571

Some format issues inherent in the e-media version may also appear in this print version.

Printed by Curran Associates, Inc. (2016)

For permission requests, please contact the Institute of Physics at the address below.

Institute of Physics
Dirac House, Temple Back
Bristol BS1 6BE UK

Phone: 44 1 17 929 7481
Fax: 44 1 17 920 0979

techtracking@iop.org

Additional copies of this publication are available from:

Curran Associates, Inc.
57 Morehouse Lane
Red Hook, NY 12571 USA
Phone: 845-758-0400
Fax: 845-758-2634
Email: curran@proceedings.com
Web: www.proceedings.com

Table of contents

Volume 657

5th International Workshop on New Computational Methods for Inverse Problems (NCMIP2015)
May 29, 2015, Cachan, France

Accepted papers received: 28 October 2015
Published online: 16 November 2015

Foreword

Papers

Journal of Physics: Conference Series **657** (2015) 011001 doi:10.1088/1742-6596/657/1/011001

Foreword

This volume of *Journal of Physics: Conference Series* is dedicated to the scientific research presented during the 5th International Workshop on New Computational Methods for Inverse Problems, NCMIP 2015 (http://complement.farman.ens-cachan.fr/NCMIP_2015.html). This workshop took place at Ecole Normale Supérieure de Cachan, on May 29, 2015. The prior editions of NCMIP also took place in Cachan, France, firstly within the scope of ValueTools Conference, in May 2011, and secondly at the initiative of Institut Farman, in May 2012, May 2013 and May 2014.

The New Computational Methods for Inverse Problems (NCMIP) workshop focused on recent advances in the resolution of inverse problems. Inverse problems appear in numerous scientific disciplines such as geophysics, biological and medical imaging, material and structure characterization, electrical, mechanical and civil engineering, and finances. The resolution of inverse problems consists of estimating the parameters of the observed system or structure from data collected by an instrumental sensing or imaging device. Its success firstly requires the collection of relevant observation data. It also requires accurate models describing the physical interactions between the instrumental device and the observed system, as well as the intrinsic properties of the solution itself. Finally, it requires the design of robust, accurate and efficient inversion algorithms. Advanced sensor arrays and imaging devices provide high rate and high volume data; in this context, the efficient resolution of the inverse problem requires the joint development of new models and inversion methods, taking computational and implementation aspects into account. During this one-day workshop, researchers had the opportunity to bring to light and share new techniques and results in the field of inverse problems.

The topics of the workshop were: algorithms and computational aspects of inversion, Bayesian estimation, Kernel methods, learning methods, convex optimization, free discontinuity problems, metamodels, proper orthogonal decomposition, reduced models for the inversion, non-linear inverse scattering, image reconstruction and restoration, and applications (bio-medical imaging, non-destructive evaluation...).

NCMIP 2015 was a one-day workshop held in May 2015 which attracted around 70 attendees. Each of the submitted papers has been reviewed by two reviewers. There are 15 accepted papers. In addition, three international speakers were invited to present a longer talk.

The workshop was supported by Institut Farman (ENS Cachan, CNRS) and endorsed by the following French research networks: GDR ISIS, GDR MIA, GDR MOA and GDR Ondes. The program committee acknowledges the following research laboratories: CMLA, LMT, LURPA and SATIE.

Eric Vourc'h and Thomas Rodet

Journal of Physics: Conference Series **657** (2015) 011001 doi:10.1088/1742-6596/657/1/011001

Workshop co-chairs:

Eric Vourc'h, SATIE laboratory, Ecole Normale Supérieure de Cachan, CNRS, France
Thomas Rodet, SATIE laboratory, Ecole Normale Supérieure de Cachan, CNRS, France

Technical program committee:

Alexandre Baussard, ENSTA Bretagne, Lab-STICC, France
Marc Bonnet, ENSTA, ParisTech, France
Laure Blanc-Féraud, I3S laboratory and INRIA Nice Sophia-Antipolis, France
Antonin Chambolle, CMAP, Ecole Polytechnique, CNRS, France
Oliver Dorn, School of Mathematics, University of Manchester, UK
Cécile Durieu, SATIE, ENS Cachan, CNRS, France
Laurent Fribourg, LSV, ENS Cachan, CNRS, France
Jerôme Idier, IRCCyN Laboratory, Ecole Centrale de Nantes, France
Pierre-Yves Joubert, IEF, Paris-Sud University, CNRS, France
Marc Lambert, L2S Laboratory, CNRS, SupElec, Paris-Sud University, France
Dominique Lesselier, L2S Laboratory, CNRS, SupElec, Paris-Sud University, France
Matteo Pastorino, DIBE, University of Genoa, Italy
Gabriel Peyré, Ceremade laboratory, University of Paris Dauphine, France
Anthony Quinn, Trinity College Dublin, Ireland
Christian Rey, LMT, ENS Cachan, CNRS, France
Jean-Yves Tourneret, IRIT laboratory, ENSEEIHT (Toulouse), France

Local chairs:

Virginie Pauchont, CMLA Laboratory, ENS Cachan, Paris-Sud University, CNRS, France
Sophie Abriet, SATIE Laboratory, ENS Cachan, Paris-Sud University, CNRS, France
Béatrice Bacquet, SATIE Laboratory, ENS Cachan, Paris-Sud University, CNRS, France

Invited speakers :

Habib Ammari, Centre de Mathématiques Appliquées (CMAP Ecole Polytechnique-CNRS), France.
Emmanuelle Gouillart, Saint-Gobain Surface du Verre et Interfaces laboratory (CNRS), France.
Otmar Scherzer, Johann Radon Institute for Computational and Applied Mathematics (RICAM), University of Vienna, Austria.

Journal of Physics: Conference Series **657** (2015) 011002 doi:10.1088/1742-6596/657/1/011002

Peer review statement

All papers published in this volume of *Journal of Physics: Conference Series* have been peer reviewed through processes administered by the proceedings Editors. Reviews were conducted by expert referees to the professional and scientific standards expected of a proceedings journal published by IOP Publishing.

Far field imaging of a dielectric inclusion

Abdul Wahab[1], **Naveed Ahmed**[2], **Tasawar Abbas**[3]

[1]Department of Mathematics, COMSATS Institute of Information Technology, 47040, Wah Cantt., Pakistan

[2]Weierstrass Institute for Applied Analysis and Stochastics, Leibniz Institute in Forschungsverbund Berlin e. V. (WIAS), Mohrenstr. 39, 10117 Berlin, Germany

[3]Department of Mathematics & Statistics, FBAS, International Islamic University, 44000, Islamabad, Pakistan

E-mail: Wahab@ciitwah.edu.pk, Naveed.Ahmed@wias-berlin.de, Tasawar44@hotmail.com

Abstract. A non-iterative topological sensitivity framework for guaranteed far field detection of a dielectric inclusion is presented. The cases of single and multiple measurements of the electric far field scattering amplitude at a fixed frequency are taken into account. The performance of the algorithm is analyzed theoretically in terms of resolution, stability, and signal-to-noise ratio.

1. Introduction

A thriving interest has been observed in topological sensitivity frameworks to procure solutions of assorted inverse problems especially for detecting small inhomogeneities and cracks embedded in homogeneous media [2, 7, 9, 10]. The impetus behind this curiosity is the robustness and simplicity of these algorithms. However, the understanding and utilization was somehow heuristic. For the first time, Ammari et al. rigorously debated these algorithms [1, 3, 11].

Our aim here is to design and debate a far field detection algorithm based on topological sensitivity of small dielectric inclusion $D_\rho := \mathbf{z}_D + \rho O_D \subset \mathbb{R}^3$ with position $\mathbf{z}_D$, scale factor ρ and a smooth bounded domain $O_D \subset \mathbb{R}^3$. The inclusion D_ρ (with permittivity $\epsilon_1 = \epsilon_0\epsilon_{1r} > 0$ and permeability $\mu_0 > 0$) is assumed to be embedded in $\mathbb{R}^3$ (with background permittivity $\epsilon_0 > 0$ and permeability $\mu_0 > 0$, setting forth a non-permeable behavior (but the results extend to the case otherwise). Here $\epsilon_{1r} > 0$ is the relative permittivity. The medium is probed by the incident electric field

$$\mathbf{E}_0(\mathbf{x}) := \nabla_{\mathbf{x}} \times (\theta^{\perp} e^{i\kappa\theta^T \mathbf{x}}) = i\kappa\theta \times \theta^{\perp} e^{i\kappa\theta^T \mathbf{x}}, \qquad \mathbf{x} \in \mathbb{R}^3, \tag{1}$$

where $\kappa = \omega\sqrt{\epsilon_0\mu_0} \in \mathbb{R}_+$ is the wavenumber, $\omega > 0$ is the angular frequency, $\theta \in \mathbb{S}^2 := \{\mathbf{x} \in \mathbb{R}^3 \,|\, \mathbf{x}\cdot\mathbf{x} = 1\}$ and $\theta^{\perp}$ is any vector orthogonal to θ. Then, the *total* electric field $\mathbf{E}_\rho$ in $\mathbb{R}^3$ in the presence of D_ρ satisfies

$$\nabla \times \nabla \times \mathbf{E}_\rho(\mathbf{x}) - \kappa^2(1 - (1-\epsilon_{1r})\chi_{D_\rho}(\mathbf{x}))\mathbf{E}_\rho(\mathbf{x}) = \mathbf{0}, \qquad \mathbf{x} \in \mathbb{R}^3, \tag{2}$$

$$\lim_{|\mathbf{x}|\to\infty} |\mathbf{x}| \left[\nabla \times (\mathbf{E}_\rho - \mathbf{E}_0) \times \hat{\mathbf{x}} - i\kappa (\mathbf{E}_\rho - \mathbf{E}_0)\right] = \mathbf{0}, \tag{3}$$

where $\chi_{D_\rho}(\mathbf{x})$ is the characteristic function of D_ρ and $\widehat{\mathbf{v}} := \mathbf{v}/|\mathbf{v}|$ for all $\mathbf{v} \in \mathbb{R}^3 \setminus \{\mathbf{0}\}$. Let $\mathbf{E}_\rho^\infty \in L_T^2(\mathbb{S}^2) := \{\hat{\mathbf{v}} \in L^2(\mathbb{S}^2) \,|\, \mathbf{v}(\hat{\mathbf{x}}) \cdot \hat{\mathbf{x}} = 0, \forall \hat{\mathbf{x}} \in \mathbb{S}^2\}$ (called *far-field amplitude*) be defined by

$$\mathbf{E}_\rho(\mathbf{x}) - \mathbf{E}_0(\mathbf{x}) = \mathbf{E}_\rho^\infty\left(\hat{\mathbf{x}}; \theta, \theta^\perp\right) e^{i\kappa|\mathbf{x}|}/|\mathbf{x}| + O\left(1/|\mathbf{x}|^2\right), \quad \text{as } |\mathbf{x}| \to \infty. \tag{4}$$

Then, the following problems are dealt with henceforth.

(i) Given the single data set $\{\mathbf{E}_\rho^\infty(\hat{\mathbf{x}}; \theta), \forall \hat{\mathbf{x}} \in \mathbb{S}^2, \theta \in \mathbb{S}^2\}$, find location $\mathbf{z}_D$ of inclusion D_ρ.

(ii) Given the multiple data set $\{\mathbf{E}_\rho^\infty(\hat{\mathbf{x}}; \theta_j), \forall \hat{\mathbf{x}} \in \mathbb{S}^2, (\theta_j)_{j=1}^n \in \mathbb{S}^2, n \in \mathbb{N}\}$, find location $\mathbf{z}_D$.

In Section 2, a topological sensitivity based location search algorithm is presented and its resolution is ascertained. A stability analysis is performed in Section 3.

2. Topological sensitivity framework and resolution analysis

Consider a search point $\mathbf{z}_S \in \mathbb{R}^3$ and create a trial inclusion $D_\delta = \mathbf{z}_S + \delta O_S$ inside $\mathbb{R}^3$ with permittivity $\epsilon_2 = \epsilon_0\epsilon_{2r} > 0$, scale factor $\delta > 0$ and smooth domain $O_S \subset \mathbb{R}^3$. Let $\mathbf{E}_\delta$ be the electric field in $\mathbb{R}^3$ in the presence of D_δ subject to incidence $\mathbf{E}_0$ and satisfying (2)-(3) with $(\epsilon_{1r}, \chi_{D_\rho}, D_\rho)$ replaced with $(\epsilon_{2r}, \chi_{D_\delta}, D_\delta)$. Let $\mathbf{E}_\delta^\infty$ be the far field amplitude of $\mathbf{E}_\delta$.

Define the misfit $\mathcal{J}[\mathbf{E}_0]$ and its topological derivative (imaging) functional $\partial_T\mathcal{J}[\mathbf{E}_0]$ by

$$\mathcal{J}[\mathbf{E}_0](\mathbf{z}_S) := \frac{1}{2}\int_{\mathbb{S}^2} \left|\mathbf{E}_\rho^\infty(\hat{\mathbf{x}}; \theta, \theta^\perp) - \mathbf{E}_\delta^\infty(\hat{\mathbf{x}}; \theta, \theta^\perp)\right|^2 ds(\hat{\mathbf{x}}), \quad \partial_T\mathcal{J}[\mathbf{E}_0](\mathbf{z}_S) := -\frac{\partial\mathcal{J}[\mathbf{E}_0](\mathbf{z}_S)}{\partial(\delta)^3}. \tag{5}$$

It is claimed that $\mathbf{z}_S \in \mathbb{R}^3$ relative to which $\partial_T\mathcal{J}[\mathbf{E}_0]$ defined in (5) has a sharp peak is a candidate for $\mathbf{z}_D$. The claim is justified in the sequel. Recall first the asymptotic expansion of $\mathbf{E}_\rho^\infty$ versus ρ [5]:

$$\mathbf{E}_\rho^\infty(\hat{\mathbf{x}}; \theta) = -\kappa^2\rho^3(4\pi)^{-1}\left(\epsilon_{1r}^{-1} - 1\right)\left(\mathbf{I} - \hat{\mathbf{x}}\hat{\mathbf{x}}^T\right)\mathbf{M}_\rho\mathbf{E}_0(\mathbf{z}_D)e^{-i\kappa\hat{\mathbf{x}}^T\mathbf{z}_D} + O(\rho^4), \tag{6}$$

wherein $\mathbf{M}_\rho = \mathbf{M}_\rho(\epsilon_{1r}^{-1}, O_D)$ is the polarization tensor. Note that $\mathbf{M}_\rho = 3\epsilon_{1r}(\epsilon_{1r} + 2)^{-1}|O_D|\mathbf{I}$ for a spherical domain, where $\mathbf{I}$ is the 3×3 identity matrix and superposed T indicates matrix transpose. It is worthwhile precising that $\mathbf{E}_\delta^\infty$ also admits an analogous expansion versus δ. Further, consider the dyadic Green's function

$$\mathbf{\Gamma}(\mathbf{x}, \mathbf{y}) := -\epsilon_0\left[\mathbf{I} + \kappa^{-2}\nabla_\mathbf{x}\nabla_\mathbf{x}^T\right]e^{i\kappa|\mathbf{x}-\mathbf{y}|}/4\pi|\mathbf{x} - \mathbf{y}|, \tag{7}$$

satisfying $\nabla_\mathbf{x} \times \nabla_\mathbf{x} \times \mathbf{\Gamma}(\mathbf{x}, \mathbf{y}) - \kappa^2\mathbf{\Gamma}(\mathbf{x}, \mathbf{y}) = -\epsilon_0\delta_\mathbf{y}(\mathbf{x})\mathbf{I}$ for all $\mathbf{x}, \mathbf{y} \in \mathbb{R}^3$ subject to the *Silver-Müller* condition. Here $\delta_\mathbf{y}(\cdot) = \delta_0(\cdot - \mathbf{y})$ is the Dirac mass at $\mathbf{y}$ and $\nabla\times$ acts on matrices column-wise. Moreover, for isotropic dielectric materials (see, for example, [4])

$$\mathbf{\Gamma}(\mathbf{x}, \mathbf{y}) = \mathbf{\Gamma}(\mathbf{y}, \mathbf{x}) \quad \text{and} \quad \nabla_\mathbf{x} \times \mathbf{\Gamma}(\mathbf{x}, \mathbf{y}) = \left[\nabla_\mathbf{y} \times \mathbf{\Gamma}(\mathbf{y}, \mathbf{x})\right]^T, \quad \forall \mathbf{x}, \mathbf{y} \in \mathbb{R}^3. \tag{8}$$

Finally, recall that the electric *Herglotz* wave for all $\Phi \in L_t^2(\mathbb{S}^2)$ is defined by ([8, Ch. 6])

$$\mathcal{H}_E[\Phi](\mathbf{z}) := \int_{\mathbb{S}^2} \Phi(\hat{\mathbf{x}})e^{i\kappa\hat{\mathbf{x}}^T\mathbf{z}}ds(\hat{\mathbf{x}}), \quad \mathbf{z} \in \mathbb{R}^3. \tag{9}$$

Lemma 2.1. *For all search points* $\mathbf{z}_S \in \mathbb{R}^3$

$$\partial_T\mathcal{J}[\mathbf{E}_0](\mathbf{z}_S) = -\kappa^2(4\pi)^{-1}\left(\epsilon_{2r}^{-1} - 1\right)\Re e\left\{\overline{\mathcal{H}_E[\mathbf{E}_\rho^\infty(\cdot; \theta)](\mathbf{z}_S)} \cdot \mathbf{M}_\delta\mathbf{E}_0(\mathbf{z}_S)\right\}. \tag{10}$$

Proof. Expanding misfit (5) and invoking expansion (6) for $\mathbf{E}_\delta^\infty$ render

$$\begin{aligned}\mathcal{J}[\mathbf{E}_0](\mathbf{z}_S) - \frac{1}{2}\int_{\mathbb{S}^2}\left|\mathbf{E}_\rho^\infty(\hat{\mathbf{x}};\theta)\right|^2 ds(\hat{\mathbf{x}}) &= -\Re e\left\{\int_{\mathbb{S}^2}\mathbf{E}_\delta^\infty(\hat{\mathbf{x}};\theta)\cdot\overline{\mathbf{E}_\rho^\infty(\hat{\mathbf{x}};\theta)}ds(\hat{\mathbf{x}})\right\} + O(\delta^6),\\ &= \frac{\kappa^2\delta^3}{4\pi}\left(\epsilon_{2r}^{-1}-1\right)\Re e\left\{\int_{\mathbb{S}^2}\left(\mathbf{I}-\hat{\mathbf{x}}\hat{\mathbf{x}}^T\right)\mathbf{M}_\delta\mathbf{E}_0(\mathbf{z}_D)\cdot\overline{\mathbf{E}_\rho^\infty(\hat{\mathbf{x}};\theta)}e^{-i\kappa\hat{\mathbf{x}}^T\mathbf{z}_S}ds(\hat{\mathbf{x}})\right\} + O(\delta^4).\end{aligned} \tag{11}$$

Remark that for any matrix $\mathbf{A}$, vectors $\mathbf{u},\mathbf{v}$, density $\Phi\in L_t^2(\mathbb{S}^2)$ and $\mathbf{x},\mathbf{z}\in\mathbb{R}^3$

$$\mathbf{A}\mathbf{u}\cdot\mathbf{v} = \mathbf{A}^T\mathbf{v}\cdot\mathbf{u}, \qquad \left(\mathbf{I}-\hat{\mathbf{x}}\hat{\mathbf{x}}^T\right)\mathbf{u} = -\hat{\mathbf{x}}\times(\hat{\mathbf{x}}\times\mathbf{u}) \tag{12}$$

$$\kappa^2\hat{\mathbf{x}}\times\left(\hat{\mathbf{x}}\times\overline{\Phi(\hat{\mathbf{x}})}\right)e^{-i\kappa\hat{\mathbf{x}}^T\mathbf{z}} = -\nabla_{\mathbf{z}}\times\nabla_{\mathbf{z}}\times\left(\overline{\Phi(\hat{\mathbf{x}})e^{i\kappa\hat{\mathbf{x}}^T\mathbf{z}}}\right). \tag{13}$$

Therefore, by the fact that $\nabla\times\nabla\times\mathcal{H}_E[\Phi]-\kappa^2\mathcal{H}_E[\Phi]=\mathbf{0}$ for $\Phi\in L_t^2(\mathbb{S}^2)$ and by above identities

$$\mathcal{J}[\mathbf{E}_0](\mathbf{z}_S) - \frac{1}{2}\int_{\mathbb{S}^2}\left|\mathbf{E}_\rho^\infty(\hat{\mathbf{x}};\theta)\right|^2 ds(\hat{\mathbf{x}}) = \frac{\kappa^2\delta^3}{4\pi}\left(\epsilon_{2r}^{-1}-1\right)\Re e\left\{\overline{\mathcal{H}_E[\mathbf{E}_\rho^\infty](\mathbf{z}_S)}\cdot\mathbf{M}_\delta\mathbf{E}_0(\mathbf{z}_S)\right\} + O(\delta^4).$$

Finally, the proof is completed by taking limit $\delta^3\to 0$. □

In order to substantiate the capability of $\partial_T\mathcal{J}[\mathbf{E}_0]$ to locate $\mathbf{z}_D$, we elaborate $\mathcal{H}_E[\mathbf{E}_\rho^\infty](\mathbf{z})$ as

$$\begin{aligned}\mathcal{H}_E[\mathbf{E}_\rho^\infty(\cdot;\theta)](\mathbf{z}) &= \int_{\mathbb{S}^2}\mathbf{E}_\rho^\infty(\hat{\mathbf{x}};\theta)e^{i\kappa\hat{\mathbf{x}}^T\mathbf{z}}ds(\hat{\mathbf{x}}),\\ &= -\frac{\kappa^2\rho^3}{4\pi}\left(\epsilon_{1r}^{-1}-1\right)\int_{\mathbb{S}^2}\left(\mathbf{I}-\hat{\mathbf{x}}\hat{\mathbf{x}}^T\right)e^{i\kappa\hat{\mathbf{x}}^T(\mathbf{z}-\mathbf{z}_D)}\mathbf{M}_\rho\mathbf{E}_0(\mathbf{z}_D)ds(\hat{\mathbf{x}}) + O(\rho^4),\\ &= -\frac{\kappa^2\rho^3}{4\pi}\left(\epsilon_{1r}^{-1}-1\right)\left[\left(\mathbf{I}-\frac{1}{\kappa^2}\nabla_{\mathbf{z}}\nabla_{\mathbf{z}}^T\right)\int_{\mathbb{S}^2}e^{i\kappa\hat{\mathbf{x}}^T(\mathbf{z}-\mathbf{z}_D)}ds(\hat{\mathbf{x}})\right]\mathbf{M}_\rho\mathbf{E}_0(\mathbf{z}_D) + O(\rho^4)\\ &= \kappa\rho^3\epsilon_0^{-1}\left(\epsilon_{1r}^{-1}-1\right)\Im m\left\{\mathbf{\Gamma}(\mathbf{z},\mathbf{z}_D)\right\}\mathbf{M}_\rho\mathbf{E}_0(\mathbf{z}_D) + O(\rho^4).\end{aligned} \tag{14}$$

Indeed, the last identity is evident since

$$\int_{\mathbb{S}^2}e^{i\kappa\hat{\mathbf{x}}^T(\mathbf{z}-\mathbf{z}_D)}ds(\hat{\mathbf{x}}) = j_0(\kappa|\mathbf{z}-\mathbf{z}_D|) = 4\pi\kappa^{-1}\Im m\left\{e^{i\kappa|\mathbf{z}-\mathbf{z}_D|}/4\pi|\mathbf{z}-\mathbf{z}_D|\right\}, \tag{15}$$

where j_n is the order n spherical Bessel function of first kind and consequently,

$$\left(\mathbf{I}-\frac{1}{\kappa^2}\nabla_{\mathbf{z}}\nabla_{\mathbf{z}}^T\right)\int_{\mathbb{S}^2}e^{i\kappa\hat{\mathbf{x}}^T(\mathbf{z}-\mathbf{z}_D)}ds(\hat{\mathbf{x}}) = -\frac{4\pi}{\kappa\epsilon_0}\Im m\left\{\mathbf{\Gamma}(\mathbf{z},\mathbf{z}_D)\right\}. \tag{16}$$

In fact, the following result is proved by virtue of Lemma 2.1 and expansion (14).

Theorem 2.2. *For all* $\mathbf{z}_S\in\mathbb{R}^3$ *and constant* $C_\epsilon := \left(\epsilon_{1r}^{-1}-1\right)\left(\epsilon_{2r}^{-1}-1\right)$

$$\partial_T\mathcal{J}[\mathbf{E}_0](\mathbf{z}_S) = -\rho^3\kappa^3C_\epsilon(4\pi\epsilon_0)^{-1}\Re e\left\{\Im m\left\{\mathbf{\Gamma}(\mathbf{z}_S,\mathbf{z}_D)\right\}\mathbf{M}_\rho\overline{\mathbf{E}_0(\mathbf{z}_D)}\cdot\mathbf{M}_\delta\mathbf{E}_0(\mathbf{z}_S)\right\} + O(\rho^4). \tag{17}$$

Remark 2.3. *It immediately follows from Theorem 2.2 that*

$$\partial_T\mathcal{J}[\mathbf{E}_0](\mathbf{z}_S) \propto \Im m\left\{\mathbf{\Gamma}(\mathbf{z}_S,\mathbf{z}_D)\right\}\mathbf{p}\cdot\mathbf{q} = -\frac{\epsilon_0\kappa}{4\pi}\left[\frac{2}{3}j_0(\kappa r)\mathbf{I} + j_2(\kappa r)\left(\hat{\mathbf{r}}\hat{\mathbf{r}}^T-\frac{1}{3}\mathbf{I}\right)\right]\mathbf{p}\cdot\mathbf{q},$$

where $\mathbf{r} := \mathbf{z}_S - \mathbf{z}_D$ *with* $r := |\mathbf{r}|$, $\mathbf{p} = \mathbf{M}_\rho \mathbf{E}_0(\mathbf{z}_D)$ *and* $\mathbf{q} = \mathbf{M}_\delta \mathbf{E}_0(\mathbf{z}_S)$. *Since* $j_n(kr) = O(1/kr)$ *as* $kr \to \infty$ *and* $j_n(kr) = O((kr)^n)$ *as* $kr \to 0$, *the functional* $\mathbf{z}_S \to \partial_T \mathcal{J}[\mathbf{E}_0](\mathbf{z}_S)$ *rapidly decays for* $\mathbf{z}_S$ *away from* $\mathbf{z}_D$ *and has a sharp peak when* $\mathbf{z}_S \to \mathbf{z}_D$ *with a focal spot size of half a wavelength of the incident wave. Moreover, it synthesizes the sensitivity of* $\mathcal{J}[\mathbf{E}_0](\mathbf{z}_S)$ *relative to the insertion of an inclusion at* $\mathbf{z}_S \in \mathbb{R}^3$. *Heuristically, if the contrast* $(\epsilon_{2r} - 1)$ *has the same sign as* $(\epsilon_{1r} - 1)$, *then* $\mathcal{J}[\mathbf{E}_0](\mathbf{z}_S)$ *must observe the most pronounced decrease at the potential candidate* $\mathbf{z}_S \in \Omega$ *for the true location* $\mathbf{z}_D$. *In other words,* $\mathbf{z}_S \to \frac{\partial \mathcal{J}[\mathbf{E}_0]}{\partial(\delta^3)}(\mathbf{z}_S)$ *is expected to attain its most pronounced negative value; refer, for instance, to [6, 7] for detailed discussions on sign heuristic. Notice that* C_ϵ *is positive if the contrasts of true and trial inclusions have same signs. Consequently, by virtue of the decay property,* $\partial_T \mathcal{J}[\mathbf{E}_0](\mathbf{z}_S) = -\frac{\partial \mathcal{J}[\mathbf{E}_0]}{\partial(\delta^3)}(\mathbf{z}_S)$ *assumes its maximum positive value when* $\mathbf{z}_S \to \mathbf{z}_D$. *In a nutshell,* $\partial_T \mathcal{J}[\mathbf{E}_0]$ *achieves the detection of* $\mathbf{z}_D$ *and resolution limit.*

2.1. Imaging with multiple measurement

Let $(\theta_j)_{j=1}^n \in \mathbb{S}^2$ be equidistributed directions and $\theta_j^{\perp,\ell}$ $(\ell = 1, 2)$ be such that $\{\theta_j, \theta_j^{\perp,1}, \theta_j^{\perp,2}\}$ forms an orthonormal basis of $\mathbb{R}^3$. We define the incident fields and the topological sensitivity functional by

$$\mathbf{E}_0^{j,\ell} := i\kappa\theta_j \times \theta_j^{\perp,\ell} e^{i\kappa\theta_j^T \mathbf{x}} \quad \text{and} \quad \partial_T \mathcal{J}(\mathbf{z}_S) := \frac{1}{n}\sum_{\ell=1}^{2}\sum_{j=1}^{n} \partial_T \mathcal{J}[\mathbf{E}_0^{j,\ell}](\mathbf{z}_S). \tag{18}$$

As for n sufficiently large $\frac{1}{n}\sum_{j=1}^n e^{i\kappa\theta_j^T(\mathbf{x}-\mathbf{y})} \approx j_0(\kappa|\mathbf{x}-\mathbf{y}|)$ and $\{\theta_j, \theta_j^{\perp,1}, \theta_j^{\perp,2}\}$ is a basis of $\mathbb{R}^3$, therefore

$$\begin{aligned}\frac{1}{n}\sum_{\ell=1}^{2}\sum_{j=1}^{n} e^{i\kappa\theta_j^T(\mathbf{x}-\mathbf{y})}\theta_j^{\perp,\ell}\left(\theta_j^{\perp,\ell}\right)^T &= \frac{1}{n}\sum_{j=1}^{n}\left(\mathbf{I} - \theta_j\theta_j^T\right) e^{i\kappa\theta_j^T(\mathbf{x}-\mathbf{y})} \\ &= \frac{1}{n}\sum_{j=1}^{n}\left(\mathbf{I} - \frac{1}{\kappa^2}\nabla_\mathbf{x}\nabla_\mathbf{x}^T\right) e^{i\kappa\theta_j^T(\mathbf{x}-\mathbf{y})} \approx -4\pi(\kappa\epsilon_0)^{-1}\Im m\{\mathbf{\Gamma}(\mathbf{x},\mathbf{y})\}.\end{aligned}$$

Similarly,

$$\frac{1}{n}\sum_{\ell=1}^{2}\sum_{j=1}^{n} e^{i\kappa\theta_j^T(\mathbf{x}-\mathbf{y})}\left(\theta_j \times \theta_j^{\perp,\ell}\right)\left(\theta_j \times \theta_j^{\perp,\ell}\right)^T \approx -4\pi(\kappa\epsilon_0)^{-1}\Im m\{\mathbf{\Gamma}(\mathbf{x},\mathbf{y})\}. \tag{19}$$

Following result holds.

Theorem 2.4. *Let* $\mathbf{z}_S \in \mathbb{R}^3$, $n \in \mathbb{N}$ *be sufficiently large and* $\mathbf{A} : \mathbf{B} = \sum_{i,j=1}^3 \mathbf{A}_{ij}\mathbf{B}_{ij}$. *Then*

$$\partial_T \mathcal{J}(\mathbf{z}_S) \approx \rho^3\kappa^4\epsilon_0^{-2}C_\epsilon \Re e\Big\{\Im m\left\{\mathbf{\Gamma}(\mathbf{z}_S,\mathbf{z}_D)\right\}\mathbf{M}_\rho : \mathbf{M}_\delta \Im m\left\{\mathbf{\Gamma}(\mathbf{z}_S,\mathbf{z}_D)\right\}\Big\} + O(\rho^4). \tag{20}$$

Proof. Since $\mathbf{A}\theta \cdot \theta = \mathbf{A} : \theta\theta^T$, the approximation (19) yields

$$\begin{aligned}\partial_T \mathcal{J}(\mathbf{z}_S) &= -\frac{\rho^3\kappa^3 C_\epsilon}{4\pi\epsilon_0 n}\sum_{\ell=1}^{2}\sum_{j=1}^{n}\Re e\Big\{\Im m\left\{\mathbf{\Gamma}(\mathbf{z}_S,\mathbf{z}_D)\right\}\mathbf{M}_\rho\overline{\mathbf{E}_0^{j,\ell}(\mathbf{z}_D)} \cdot \mathbf{M}_\delta \mathbf{E}_0^{j,\ell}(\mathbf{z}_S)\Big\} + O(\rho^4), \\ &= -\frac{\rho^3\kappa^5 C_\epsilon}{4\pi\epsilon_0 n}\sum_{\ell=1}^{2}\sum_{j=1}^{n}\Re e\Big\{\Im m\left\{\mathbf{\Gamma}(\mathbf{z}_S,\mathbf{z}_D)\right\}\mathbf{M}_\rho : \mathbf{M}_\delta(\theta_j \times \theta_j^{\perp,\ell})(\theta_j \times \theta_j^{\perp,\ell})^T e^{i\kappa\theta_j^T(\mathbf{z}_S-\mathbf{z}_D)}\Big\} + O(\rho^4), \\ &\approx \rho^3\kappa^4\epsilon_0^{-2}C_\epsilon\Re e\{\Im m\left\{\mathbf{\Gamma}(\mathbf{z}_S,\mathbf{z}_D)\right\}\mathbf{M}_\rho : \mathbf{M}_\delta\Im m\left\{\mathbf{\Gamma}(\mathbf{z}_S,\mathbf{z}_D)\right\}\} + O(\rho^4).\end{aligned}$$

□

Theorem 2.4 substantiates that $\partial_T\mathcal{J}(\mathbf{z}_S) \propto \Im m\left\{\mathbf{\Gamma}(\mathbf{z}_S,\mathbf{z}_D)\right\} : \Im m\left\{\mathbf{\Gamma}(\mathbf{z}_S,\mathbf{z}_D)\right\}$ and has a peak (sharper than that of $\partial_T\mathcal{J}[\mathbf{E}_0]$) when $\mathbf{z}_S \to \mathbf{z}_D$ (see Remark 2.3).

3. Stability with respect to measurement noise

The aim in this section is to establish stability of the imaging functional $\partial_T\mathcal{J}$. Assume that the measurements of the far field amplitude are corrupted by a mean-zero circular Gaussian noise $\boldsymbol{\xi} : \mathbb{S}^2 \to \mathbb{C}^3$, that is, $\mathbf{E}^\infty_\rho(\hat{\mathbf{x}}) := \widetilde{\mathbf{E}}^\infty_\rho(\hat{\mathbf{x}}) + \boldsymbol{\xi}(\hat{\mathbf{x}})$ for all $\hat{\mathbf{x}} \in \mathbb{S}^2$, where $\mathbf{E}^\infty_\rho$ is the corrupted value. Henceforth, a superposed $\sim$ indicates true value without noise corruption. Further, let σ_ξ be the noise covariance of $\boldsymbol{\xi}$ such that for all $\hat{\mathbf{y}}, \hat{\mathbf{y}}' \in \mathbb{S}^2$

$$\mathbb{E}\left[\boldsymbol{\xi}(\hat{\mathbf{y}})\overline{\boldsymbol{\xi}(\hat{\mathbf{y}}')}^T\right] = {\sigma_\xi}^2\delta_{\hat{\mathbf{y}}}(\hat{\mathbf{y}}')\mathbf{I} \qquad \text{and} \qquad \mathbb{E}\left[\boldsymbol{\xi}^j(\hat{\mathbf{y}})\overline{\boldsymbol{\xi}^{j'}(\hat{\mathbf{y}}')}^T\right] = {\sigma_\xi}^2\delta_{jj'}\delta_{\hat{\mathbf{y}}}(\hat{\mathbf{y}}')\mathbf{I}, \tag{21}$$

where j and j' indicate the j-th and j'-th measurements and $\delta_{jj'}$ is Kronecker's delta.

The noise affects $\partial_T\mathcal{J}[\mathbf{E}_0]$ and $\partial_T\mathcal{J}$ by means of Herglotz wave, $\mathcal{H}_E[\mathbf{E}^\infty_\rho] = \mathcal{H}_E[\widetilde{\mathbf{E}}^\infty_\rho] + \mathcal{H}_E[\boldsymbol{\xi}]$. The term $\mathcal{H}_E[\widetilde{\mathbf{E}}^\infty_\rho]$ is independent of noise and yields true image; as discussed earlier. On the other hand, $\mathcal{H}_E[\boldsymbol{\xi}]$ is a circular Gaussian random process with mean-zero by linearity and by definition of $\boldsymbol{\xi}$. Furthermore, for all $\hat{\mathbf{z}}, \hat{\mathbf{z}}' \in \mathbb{S}^2$

$$\begin{aligned}\mathbb{E}\left[\mathcal{H}_E[\boldsymbol{\xi}](\mathbf{z})\overline{\mathcal{H}_E[\boldsymbol{\xi}](\mathbf{z}')}^T\right] =&\mathbb{E}\left[\iint_{\mathbb{S}^2\times\mathbb{S}^2} \boldsymbol{\xi}(\hat{\mathbf{x}})\overline{\boldsymbol{\xi}(\hat{\mathbf{y}})}^T e^{i\kappa\hat{\mathbf{x}}^T\mathbf{z}}e^{-i\kappa\hat{\mathbf{y}}^T\mathbf{z}'}ds(\hat{\mathbf{x}})ds(\hat{\mathbf{y}})\right],\\ =&\mathbb{E}\left[\iint_{\mathbb{S}^2\times\mathbb{S}^2} \hat{\mathbf{x}}\times(\boldsymbol{\xi}(\hat{\mathbf{x}})\times\hat{\mathbf{x}})\overline{\boldsymbol{\xi}(\hat{\mathbf{y}})}^T e^{i\kappa\hat{\mathbf{x}}^T\mathbf{z}}e^{-i\kappa\hat{\mathbf{y}}^T\mathbf{z}'}ds(\hat{\mathbf{x}})ds(\hat{\mathbf{y}})\right].\end{aligned}$$

To get the latter identity, the fact that $\boldsymbol{\xi} \in L^2_T(\mathbb{S}^2)$ is used. By virtue of (12)-(13), one gets

$$\begin{aligned}\mathbb{E}\left[\mathcal{H}_E[\boldsymbol{\xi}](\mathbf{z})\overline{\mathcal{H}_E[\boldsymbol{\xi}](\mathbf{z}')}^T\right] =&\mathbb{E}\left[\iint_{\mathbb{S}^2\times\mathbb{S}^2} (\mathbf{I}-\hat{\mathbf{x}}\hat{\mathbf{x}}^T)\boldsymbol{\xi}(\hat{\mathbf{x}})\overline{\boldsymbol{\xi}(\hat{\mathbf{y}})}^T e^{i\kappa\hat{\mathbf{x}}^T\mathbf{z}}e^{-i\kappa\hat{\mathbf{y}}^T\mathbf{z}'}ds(\hat{\mathbf{x}})ds(\hat{\mathbf{y}})\right],\\ =&{\sigma_\xi}^2\left(\mathbf{I}-\frac{1}{\kappa^2}\nabla_{\mathbf{z}}\nabla_{\mathbf{z}}^T\right)\int_{\mathbb{S}^2} e^{i\kappa\hat{\mathbf{x}}^T(\mathbf{z}-\mathbf{z}')}ds(\hat{\mathbf{x}}) = -\frac{4\pi{\sigma_\xi}^2}{\kappa\epsilon_0}\Im m\{\mathbf{\Gamma}(\mathbf{z},\mathbf{z}')\}.\end{aligned} \tag{22}$$

Thus, $\mathcal{H}_E[\boldsymbol{\xi}]$ precipitates a speckle pattern, that is, a random cloud of hot spots having typical diameters of the order of wavelength and amplitudes of the order of $\sigma_\xi/\sqrt{\kappa}$. The covariance of corrupted image is furnished by Theorem 2.2 and expression (22) as

$$\begin{aligned}&\mathrm{Cov}\big(\partial_T\mathcal{J}(\mathbf{z}),\partial_T\mathcal{J}(\mathbf{z}')\big)\\ &=\frac{a_\epsilon^2}{n^2}\sum_{\ell,\ell'=1}^{2}\sum_{j,j'=1}^{n}\mathbb{E}\left[\Re e\left\{\overline{\mathcal{H}_E[\boldsymbol{\xi}^{j,\ell}](\mathbf{z})}\cdot\mathbf{M}_\delta\mathbf{E}_0^{j,\ell}(\mathbf{z})\right\}\Re e\left\{\overline{\mathcal{H}_E[\boldsymbol{\xi}^{j',\ell'}](\mathbf{z}')}\cdot\mathbf{M}_\delta\mathbf{E}_0^{j',\ell'}(\mathbf{z}')\right\}\right],\\ &=\frac{a_\epsilon^2}{2n^2}\sum_{\ell=1}^{2}\sum_{j=1}^{n}\Re e\left\{\mathbf{M}_\delta\mathbf{E}_0^{j,\ell}(\mathbf{z})\cdot\mathbb{E}\left[\overline{\mathcal{H}_E[\boldsymbol{\xi}^{j,\ell}](\mathbf{z})}\Big(\mathcal{H}_E[\boldsymbol{\xi}^{j,\ell}](\mathbf{z}')\Big)^T\right]\overline{\mathbf{M}_\delta\mathbf{E}_0^{j,\ell}(\mathbf{z}')}\right\},\\ &=-\frac{2\pi a_\epsilon^2{\sigma_\xi}^2}{n^2\kappa\epsilon_0}\sum_{j=1}^{n}\Re e\left\{\mathbf{M}_\delta\mathbf{E}_0^{j,\ell}(\mathbf{z})\cdot\Im m\{\mathbf{\Gamma}(\mathbf{z},\mathbf{z}')\}\overline{\mathbf{M}_\delta\mathbf{E}_0^{j,\ell}(\mathbf{z}')}\right\},\end{aligned}$$

where $a_\epsilon = \kappa^2(4\pi)^{-1}\left(\epsilon_{2r}^{-1}-1\right)$ and the identities (21)-(22) are invoked.

Journal of Physics: Conference Series **657** (2015) 012001 doi:10.1088/1742-6596/657/1/012001

Now substituting expression for $\mathbf{E}_0^{j,\ell}$ and using approximation (19), we arrive at

$$\begin{aligned}&\mathrm{Cov}\big(\partial_T\mathcal{J}(\mathbf{z}),\partial_T\mathcal{J}(\mathbf{z}')\big)\\&=-\frac{2\pi a_\epsilon^2\sigma_\xi{}^2\kappa}{n^2\epsilon_0}\sum_{\ell=1}^{2}\sum_{j=1}^{n}\Re e\Big\{\mathbf{M}_\delta(\theta_j\times\theta_j^{\perp,\ell})\cdot\Im m\big\{\mathbf{\Gamma}(\mathbf{z},\mathbf{z}')\big\}\mathbf{M}_\delta(\theta_j\times\theta_j^{\perp,\ell})e^{i\kappa\theta^T(\mathbf{z}-\mathbf{z}')}\Big\},\\&\approx\kappa^4\sigma_\xi{}^2(\epsilon_{2r}^{-1}-1)^2(2n)^{-1}\epsilon_0^{-2}\Re e\Big\{\mathbf{M}_\delta\Im m\big\{\mathbf{\Gamma}(\mathbf{z},\mathbf{z}')\big\}:\Im m\big\{\mathbf{\Gamma}(\mathbf{z},\mathbf{z}')\big\}\mathbf{M}_\delta\Big\},\end{aligned}$$

In order to fathom the statistics of corrupted image, assume for an instance that D_δ is a sphere so that $\mathbf{M}_\delta=3(2\epsilon_{2r}^{-1}+1)^{-1}|O_S|\mathbf{I}$. Then, it can be readily verified that

$$\mathrm{Cov}\big(\partial_T\mathcal{J}(\mathbf{z}),\partial_T\mathcal{J}(\mathbf{z}')\big)\approx b_\epsilon^2\sigma_\xi{}^2\kappa^4(2n)^{-1}\left\|\Im m\big\{\mathbf{\Gamma}(\mathbf{z},\mathbf{z}')\big\}\right\|^2, \tag{23}$$

where $b_\epsilon=3|O_S|(\epsilon_{2r}^{-1}-1)/\epsilon_0(2\epsilon_{2r}^{-1}+1)$ and $\|\mathbf{A}\|:=\sqrt{\mathbf{A}:\mathbf{A}}$ is the *Frobenius* norm of $\mathbf{A}$.

Remark 3.1. *The above analysis elucidates that noise perturbation of image is typically of order $\sigma_\xi/\sqrt{2n}$ and the hot spots have shape identical with that of actual peak procured by $\partial_T\mathcal{J}$ subject to accurate measurements. The main spike of $\partial_T\mathcal{J}$ is not altered. Further, the dependence of typical perturbation size on $1/\sqrt{2n}$ indicates that the topological sensitivity based imaging framework is more stable when multiple measurements are available at hand.*

3.1. Signal-to-noise ratio

It follows immediately from (23) that the variance of $\partial_T\mathcal{J}$ at $\mathbf{z}\in\mathbb{R}^3$ can be approximated by

$$\mathrm{Var}\big(\partial_T\mathcal{J}(\mathbf{z})\big)\approx b_\epsilon^2\sigma_\xi{}^2\kappa^4(2n)^{-1}\left\|\Im m\big\{\mathbf{\Gamma}(\mathbf{z},\mathbf{z})\big\}\right\|^2=\epsilon_0^2b_\epsilon^2\sigma_\xi{}^2\kappa^6(2n)^{-1}(6\pi)^{-2}. \tag{24}$$

Specifically when D_ρ is spherical inclusion, the signal-to-noise ratio (SNR) defined by SNR $=\mathbb{E}\left[\partial_T\mathcal{J}(\mathbf{z}_D)\right]\left(\mathrm{Var}\left[\partial_T\mathcal{J}(\mathbf{z}_D)\right]\right)^{-1/2}$, can be approximated by virtue of (24) and Theorem 2.4 as

$$\mathrm{SNR}\approx\sqrt{2n}(2\pi)^{-1}\sigma_\xi{}^{-1}|2\epsilon_{1r}^{-1}+1|^{-1}|\epsilon_{1r}^{-1}-1|\rho^3|O_D|\kappa^3. \tag{25}$$

The expression (25) indicates that signal-to-noise ratio depends directly on contrast, volume of the inclusion, the operating frequency and inversely on noise covariance.

4. Conclusion

The topological sensitivity functions presented herein for detecting a dielectric inclusion, from single or multiple measurements of electric far field amplitude, are proved to be very efficient in terms of resolution, stability and signal-to-noise ratio. In fact, the Rayleigh's resolution limit is achieved and stability with respect to measurement noise is substantiated. The stability with respect to medium noise is not dealt with, however, moderate stability can be guaranteed under Born approximation. The reader is referred, for instance, to [1, 11] in this regard.

References

[1] Ammari H, Bretin E, Garnier J, Jing W, Kang H and Wahab A 2013 *SIAM J. Imag. Sci.*, **6(4)** 2174–212.
[2] Ammari H, Bretin E, Garnier J, Kang H, Lee H and Wahab A 2015 *Mathematical Methods in Elasticity Imaging* (New Jersey: Princeton University Press).
[3] Ammari H, Garnier J, Jugnon V and Kang H 2012 *SIAM J. Cont. Opt.* **50(1)** 48–76.
[4] Ammari H, Iakovleva I, Lesselier D and Perrusson G 2007 *SIAM J. Sci. Comput.* **29** 674–709.
[5] Ammari H and Volkov D 2005 *Int. J. Multiscale. Comp. Eng.* **3** 149-60.
[6] Bellis C, Bonnet M, and Cakoni F 2013 *Inverse Problems*, **29** 075012.
[7] Bonnet M and Guzina B B 2004 *Internat. J. Numer. Methods Engrg.* **61** 2344–73.
[8] Colton D and Kress R 1983 *Integral Equation Methods in Scattering Theory* (New York: John Wiley & Sons).
[9] Céa J, Garreau S, Guillaume P and Masmoudi M 2001 *Comput. Methods Appl. Mech. Engrg.* **188** 703–26.
[10] Dominguez N and Gibiat V 2010, *Ultrasonics* **50** 172–79.
[11] Wahab A 2015 Stability and resolution analysis of topological derivative based localization of small electromagnetic inclusions, to appear, *SIAM Journal on Imaging Sciences*.

5th International Workshop on New Computational Methods for Inverse Problems IOP Publishing
Journal of Physics: Conference Series **657** (2015) 012002 doi:10.1088/1742-6596/657/1/012002

Parameter Identification by Iterative Constrained Regularization

Fabiana Zama
Department of Mathematics, P.zza Porta S. Donato 5, Bologna, ITALY
E-mail: fabiana.zama@unibo.it

Abstract. Parameter identification from noisy data is an ill-posed inverse problem and data noise leads to poor solutions. Regularization methods are necessary to obtain stable solutions. In this paper we introduce the regularization by means of an iteratively weighted constraint and define an algorithm to compute the weights and solve the constrained problems using as prior information the given measurements. Although this approach is general, in the present work we prove the convergence in the case of least squares data fit with ℓ_2 regularization term. The data reported in the numerical experiments prove the efficiency and good quality of the results.

1. Introduction

Parameter identification is a very important topic in applied sciences and engineering. The modeling of biological or chemical systems requires often to estimate the parameters of differential models from noisy measurements (see for example [1],[2]). Parameter identification is an optimization problem where the objective function, usually non linear, represents the fit to the observed data. The solution process requires to repeatedly solve the differential problem, called *forward problem* or *state equation*. The majority of these identification problems is ill-posed, i.e. the parameters do not depend continuously on the data, which cannot be measured exactly but are affected by noise. Therefore regularization methods have to be used in order to obtain stable solutions. Aim of this work is to present an algorithm, in a general framework, to compute a regularized solution of the parameter identification problem by adding a smoothing constraint with automatic computation of the regularization parameter. The idea of solving identification problems by a sequence of constrained optimization problems is widely present in the literature (see for example [3, 4]). Although the scheme is general, in this paper we analyze the details of choosing the ℓ_2 norm for both the objective function and the penalty term, obtaining therefore the Tikhonov regularization. The application of this regularization method to parameter identification problems has been extensively investigated in the literature [5, 6]. The main difficulty of the Tikhonov regularization is the correct computation of the regularization parameter. In our approach we define an iterative procedure to adjust the weight of the smoothing constraint, without assuming any prior knowledge of the noise level. We define the Constrained Least Squares Identification algorithm (`CLSId`), by extending the principle already used in image deblurring and denoising problems [7, 8]. The regularization method for generic objective and penalty functions is discussed in section 2, while in section 3 we report the algorithmic features for the ℓ_2 norm. The iterative procedure to compute the regularization weight is analysed is section 4. In the numerical experiments, reported in section 5, we show the results of two parameter identification test problems.

2. The Parameter Identification Algorithm

We consider the discrete finite dimensional problem of estimating the parameter $\mathbf{q} \in \mathbb{R}^K$ of a differential model (state equation) $c(\mathbf{u}, \mathbf{q}) = 0$, usually an ordinary or partial differential equation. We assume that the measurements are obtained by mapping the solution of the state equation $\mathbf{u}(\mathbf{q})$ (state variable) at some measurement points ($\mathbf{y} = F(\mathbf{u}(\mathbf{q}))$) and adding the noise, represented by a discrete normalized random distribution with zero mean and assigned variance δ:

$$\mathbf{y}^\delta = \mathbf{y} + \delta\boldsymbol{\eta} \tag{1}$$

where $\mathbf{y}, \boldsymbol{\eta} \in \mathbb{R}^N$ and $\|\boldsymbol{\eta}\| = 1$. The parameter identification is obtained by minimizing a cost function $J(\mathbf{q})$ representing the data fit of the model measurements $F(\mathbf{u}(\mathbf{q}))$ to the noisy data $\mathbf{y}^\delta$. The majority of such identification problems is ill-posed hence regularization methods have to be used in order to obtain stable solutions. In the present work we introduce a smoothing constraint $\mathcal{R}(\mathbf{q})$ and define the regularized solution $\mathbf{q}_{reg}$ as the solution of the constrained optimization problem:

$$min_{\mathbf{q}} J(\mathbf{q}), \quad \text{s.t.} \quad \mathcal{R}(\mathbf{q}) \leq \gamma \tag{2}$$

where $\gamma > 0$ defines the prescribed smoothness level required by the solution. Our aim is to define a general scheme to compute a suitable value of the regularization constraint γ and to efficiently compute the regularized solution $\mathbf{q}_{reg}$. Assuming $\mathcal{R}(\mathbf{q})$ and $J(\mathbf{q})$ are continuously differentiable and convex, problem (2) can be formulated in the equivalent Lagrangian dual form as:

$$\max_{\lambda} \Phi(\lambda), \quad \text{where} \quad \Phi(\lambda) \equiv \min_{\mathbf{q}} \mathcal{L}(\mathbf{q}, \lambda) \tag{3}$$

where $\mathcal{L}(\mathbf{q}, \lambda) \equiv J(\mathbf{q}) + \lambda\,(\mathcal{R}(\mathbf{q}) - \gamma)$.

After imposing the first order conditions $\nabla_\lambda \mathcal{L}(\mathbf{q}, \lambda) = 0$ we can define the solution $\hat{\lambda}$ of the dual problem (3) as:

$$\text{find } \hat{\lambda} \text{ s.t. } \mathcal{R}(\hat{\mathbf{q}}) - \gamma = 0, \tag{4}$$

where $\hat{\mathbf{q}}$ is the solution of the unconstrained problem:

$$\min_{\mathbf{q}} J(\mathbf{q}) + \lambda(\mathcal{R}(\mathbf{q}) - \gamma) \tag{5}$$

with $\lambda = \hat{\lambda}$. Assuming that $\mathcal{R}(\mathbf{q}(\lambda))$ is a continuously differentiable function such that

(i) $\mathcal{R}(\mathbf{q}(\lambda))$ is strictly decreasing in the interval $[\lambda_-, \lambda_+]$ $(0 \leq \lambda_- < \hat{\lambda} < \lambda_+)$.

(ii) $G(\lambda_-) > 0, \quad G(\lambda_+) < 0, \quad \gamma > 0$.

where $\mathbf{q}(\lambda) = argmin_{\mathbf{q}} \mathcal{L}(\mathbf{q}, \lambda)$ and $G(\lambda) \equiv \mathcal{R}(\mathbf{q}(\lambda)) - \gamma$, using proposition 1 in [7] we can define an hybrid bisection-secant method to compute the sequence $\{\lambda_k\}$, $k = 0, 1, \ldots$, that converges to the root $\hat{\lambda}$ of $G(\lambda)$, $\forall \lambda_0 > 0$ such that $\mathcal{R}(\mathbf{q}(\lambda_0)) - \gamma < 0$. Hence for each assigned value $\gamma > 0$, assuming that λ_0 satisfies $\mathcal{R}(\mathbf{q}(\lambda_0)) < \gamma$ and that $\mathcal{R}$ is strictly decreasing, we can compute the sequence $\{\lambda_k, \mathbf{q}_k\}$ converging to the solution of the dual problem $\hat{\lambda}$ and to the solution of the constrained problem (2) $\mathbf{q}_{reg} \equiv \hat{\mathbf{q}}$. The values $\{\lambda_k, \mathbf{q}_k\}$ are computed by means of (4), (5) as follows:

$$\lambda_k = \lambda_{k-1} + \texttt{bisection} - \texttt{secant step} \tag{6}$$

$$\mathbf{q}(\lambda_k) = argmin_{\mathbf{q}} \{J(\mathbf{q}) + \lambda_k(\mathcal{R}(\mathbf{q}) - \gamma)\} \tag{7}$$

Since the unconstrained minimization problem (5) (with $\lambda = \lambda_k$) is solved at each step of the iterations (7), then it is of great importance to use accurate and efficient methods.

3. Least Squares Regularization

The quality of the problem solution depends strongly on the specific choice of the objective function $J(\mathbf{q})$ as well as the penalty term $\mathcal{R}(\mathbf{q})$. Assuming to have Gaussian white noise on the data, the fit function J can be suitably defined by the least squares distance. In this case, in (5), we have:

$$\mathbf{q}(\lambda_k) = argmin_{\mathbf{q}} \left\{ \|F(\mathbf{q}) - \mathbf{y}^\delta\|_2^2 + \lambda_k(\mathcal{R}(\mathbf{q}) - \gamma) \right\} \tag{8}$$

A fast and simple way to solve this non linear problem is by the Gauss Newton method. Defining the Jacobian matrix $J_F(\mathbf{q}) = \partial F_i / \partial \mathbf{q}_j$, we compute the solution of (8) as $\mathbf{q}(\lambda_k) = \mathbf{p}^{(m)}$ with $\mathbf{p}^{(0)} = \mathbf{q}(\lambda_{k-1})$ and

$$\mathbf{p}^{(m)} = argmin_{\mathbf{q}} \left\{ \|J_F(\mathbf{p}^{(m-1)})(\mathbf{q} - \mathbf{p}^{(m-1)}) - \mathbf{r}^{(m-1)}\| + \lambda_k \left(\mathcal{R}(\mathbf{q}) - \gamma\right) \right\}, \quad m = 1, 2, \ldots \tag{9}$$

where $\mathbf{r}^{(m)} = \mathbf{y}^\delta - F(\mathbf{p}^{(m)})$ is the residual vector. The iteration (9) is repeated until $\|\mathbf{p}^{(m)} - \mathbf{p}^{(m-1)}\|_\infty < tol\|\mathbf{p}^{(m)}\|_\infty$. The solution $\mathbf{p}^{(m)}$ in (9) is computed by applying the Armijo damping rule, to guarantee the decrease of the regularized objective function. The parameter *tol* controls the convergence: using a too small value causes an increases of the algorithm complexity and may not represent a real improvement in the quality of the results. The computation of the residual and Jacobian matrix requires to solve the differential state equation with several values of the parameter. In the present work we solve the state equation by means of a finite difference method using central second order finite differences. The Jacobian matrix is computed by means of forward finite difference approximation. By choosing the regularization function as:

$$\mathcal{R}(\mathbf{q}) = \|L_2\mathbf{q}\|_2^2 \tag{10}$$

where L_2 is the second order difference matrix, we can prove that $\|L_2\mathbf{w}(\lambda)\|_2$ is strictly decreasing for each solution $\mathbf{w}(\lambda)$ of the minimization problem (9) with $\lambda = \lambda_k, k = 0, 1, \ldots$.

Proposition 1. *Let*

$$\pi(\lambda) = \|L_2\mathbf{w}(\lambda)\|_2, \quad \alpha = 0, 1, 2$$

where $\mathbf{w}(\lambda)$ *satisfies* (9) *for* $k = 0, 1, \ldots$ *then* $\pi(\lambda)$ *is strictly convex on* $[0, \infty)$ *and decays monotonically to zero as* λ *increases from zero.*

Proof. Applying the first order conditions to (9) we can define the solution $\mathbf{w}(\lambda)$ as:

$$\left(J_F(\mathbf{p}^{(m)})^t J_F(\mathbf{p}^{(m)}) + \lambda L_2^t L_2\right) \mathbf{w}(\lambda) = -J_F(\mathbf{p}^{(m)})^t \mathbf{z}^{(k)} \tag{11}$$

where $\mathbf{z}^{(k)} = \mathbf{r}^{(k)} - J_F(\mathbf{p}^{(m)})\mathbf{p}^{(m)}$. Introducing the Generalized Singular value Decomposition (GSVD) [9, 10] of the matrix

$$B = \begin{bmatrix} J_F(\mathbf{p}^{(m)}) \\ L_2 \end{bmatrix}, \quad \begin{matrix} J_F((\mathbf{p}^{(m)}) \in \mathbb{R}^{N \times K} \\ L_2 \in \mathbb{R}^{N_L \times K} \end{matrix}$$

in the assumption that $\text{rank}(B) = k \leq K$ and $N \geq K$:

$$J_F = UD_JX^{-1}, \quad D_j = diag(\theta_1, \ldots, \theta_K) \geq 0$$

$$L_2 = VD_LX^{-1}, \quad D_L = diag(\beta_1, \ldots, \beta_t) \geq 0$$

where $U \in \mathbb{R}^{N \times N}$ and $V \in \mathbb{R}^{N_L \times N_L}$ are orthogonal matrices and $X \in \mathbb{R}^{K \times K}$ is non singular and: $0 \leq \theta_1 \leq \theta_2 \leq \cdots \leq \theta_K \leq 1$, $1 \geq \beta_1 \geq \cdots \geq \beta_t \geq 0$, $\beta_{t+1} = \ldots = \beta_K = 0$ $t = min(N_L, K)$

Journal of Physics: Conference Series **657** (2015) 012002 doi:10.1088/1742-6596/657/1/012002

Using the GSVD in (11) we can write the solution $\mathbf{w}(\lambda)$ as :

$$\mathbf{w}(\lambda) = X^{-1}\left(D_J^t D_J + \lambda D_L^t D_L\right)^{-1} D_J^t U \mathbf{z}^{(k)}$$

and substituting it in $\pi(\lambda)$ we have:

$$\pi(\lambda) = \sqrt{(D_J X \mathbf{w}(\lambda))^t D_J X \mathbf{w}(\lambda)} = \left(\sum_{i=1}^{t} \frac{\mu_i^2}{(\mu_i^2+\lambda)^2}(U_i^t \mathbf{z}^{(k)})^2\right)^{1/2}$$

where $\mu_i = \theta_i/\beta_i$. Now we can use Lemma 2.1 in [11] defining $a_i = \mu_i U_i^t \mathbf{z}^{(k)}$, $b_i = \mu_i^2$ and

$$\pi(\lambda) = \chi(\lambda) \equiv \left(\sum_{i=1}^{t}\left(\frac{a_i}{b_i+\lambda}\right)^2\right)^{1/2}$$

and prove that $\pi(\lambda)$ is strictly convex on $[0,\infty)$ and monotonically decreasing to zero as λ increases from zero. □

This guarantees that assumption (i) at page 2 is fulfilled and the sequence $(\lambda_k, \mathbf{q}_k)$ converges to the solution of the dual problem. Hence we can define the Constrained Least Squares Identification algorithm `CLSId` as follows:

Algorithm `CLSId`(γ)

$k = 0$; $\lambda_0 = 1$,
`repeat`
    $\mathbf{q}(\lambda_k) = argmin_{\mathbf{q}}\{J(\mathbf{q}) + \lambda_k(\mathcal{R}(\mathbf{q}) - \gamma)\}$ (`Gauss Newton + Armijo`)
    $G_k = \mathcal{R}(\mathbf{q}(\lambda_k)) - \gamma$
    $\lambda_{k+1} = \lambda_k +$`hybrid bisection-secant step`
    $k = k+1$
`until` $|G_k - \gamma| < \tau_r G_0 + \tau_a$ or $|\lambda_k - \lambda_{k-1}| < \tau_a$ or $k > maxit$

where τ_a, τ_r are relative and absolute tolerance parameters.

4. Iterative estimate of the smoothing parameter γ

In order to obtain a good quality solution, a suitable value of the smoothing parameter γ needs to be computed. If γ is too small, the smoothing constraint forces the solution of (2) away from the noisy data, producing a smooth solution $\mathbf{q}$ with a large $J(\mathbf{q})$. On the other hand, if γ is too large, the regularity constraint becomes looser, allowing a better fit to the noisy data (small $J(\mathbf{q})$ value). Since the optimal value γ_{opt} lies in between these two extrema, our idea is to define a starting value $\gamma_0 < \gamma_{opt}$ and increase it on the basis of the behavior of the data fit function, with the aim to get closer to γ_{opt}. Observations performed on different problems show that $J(\mathbf{q})$ decreases steeply when $\gamma \ll \gamma_{opt}$ and tend to become almost flat when γ gets closer to γ_{opt}. Therefore we define the sequence $\gamma_0, \leq \gamma_1 \leq \cdots$ as:

$$\gamma_{\ell+1} = (1+\theta)\gamma_\ell, \;\; \text{if } J(\mathbf{q}_\ell) - J(\mathbf{q}_{\ell-1}) > \tau(\gamma_\ell - \gamma_{\ell-1}), \;\; \gamma_{\ell+1} = \gamma_\ell \text{ otherwise} \qquad (12)$$

with θ representing the relative change in $\gamma_{\ell+1}$ and τ the decrease rate of the data fit function at $\mathbf{q}_\ell$, defined as the solution of (2) with $\gamma = \gamma_\ell$. The starting value γ_0 is defined by computing a smooth approximate solution. The least squares fit function allows us to apply the Nonlinear Conjugate gradient method (NCg) to the problem $\min_{\mathbf{q}} J(\mathbf{q})$. The semi-convergence of the NCg iterations is well known and also the strong smoothness of the first NCg iterates. Using this property we compute $\mathbf{q}_{cg}$ by early stopping NCg (2-3 iterations) and define $\gamma_0 = \|L_2\mathbf{q}_{cg}\|_2^2$. Despite the number of parameter is increased, the tuning of $\tau = 10^{-p}$, $p = 1, 2$ and $\theta \in [1/4, 1/2]$ is easier compared to γ that can change of several orders of magnitudes, depending on the problem.

5. Numerical Experiments

In this section we report some numerical tests where the state equation is a second order linear boundary value problem. The experiments are carried out on Intel Core i5 1.60GHz×4 processor (5.8 GiB RAM), using Matlab 2010a. Test problems:

T1 Identification of the reaction parameter $q(x)$ in the stationary model:

$$u(x)'' - q(x)u(x) = f, \quad u(0) = u(1) = 0, \quad x \in [a,b] \equiv [0,1] \tag{13}$$

where the function f is defined by setting $u(x) = sin(\pi x)$ and the parameter $q(x) = 10(x^4 - sin(\pi x))$. The starting value $\mathbf{q}_0$ is obtained to guarantee that at least the boundary conditions are fulfilled, by sampling the function: $q_0(x) = 10(2x^3 - (1+\pi)x^2 + \pi x)$ [12].

T2 Identification of the diffusivity function $c(x)$ in the differential model:

$$(c(x)u(x)')' = f(x), \quad u(0) = u(1) = 0, \quad x \in [a,b] \equiv [0,1] \tag{14}$$

where the function f is defined by setting $u(x) = \sin(\pi x(1-x))$ and the parameter $q(x) = 1 + e^{-(10(x-0.3))^2)} + e^{-(10(x-0.7))^2)}$. The starting value $\mathbf{q}_0 = 1$ is used.

The discrete reference solution of the state equation, is computed by the finite difference method, sampling the spatial domain with a uniform grid of M points $x_i = a + (i-1)(b-a)/(M-1), \quad i = 1, \ldots, M$. The noiseless measurements $\mathbf{y}$ are obtained by interpolating the reference solution in the measurement points defined by a uniformly spaced grid of N points: $a = \xi_1 < \xi_2 < \ldots \xi_{N-1} < \xi_N = b$. Gaussian white noise, as in (1), is added with different levels $\delta \in [10^{-4}, 10^{-1}]$. The quality of the results is evaluated by the Parameter Relative Error (PRE) and the Residual Norm ($ResN$): PRE $= \|\mathbf{q}_{reg} - \mathbf{q}_{true}\|_2 / \|\mathbf{q}_{true}\|_2$, where $\mathbf{q}_{reg}$ is obtained applying the regularization algorithm `CLSId` and $\mathbf{q}_{true}$ is the discretization of the parameter $q(x)$ at the measurement points $\xi_1, \ldots, \xi_N$. The quality of the reconstructions depends on the discretization values N and M. For values of M and N fixed, the results depend on the level of noise present in the data and on the value of the smoothness level γ. To analyze the features of the regularization algorithm `CLSId`(γ) we focus on the case $M = 1001$, $N = 101$ and with tolerance parameters $\tau_a = \tau_r = 1.e-4$. The reference value γ_{cg} of the regularization constraint is obtained by stopping the nonlinear Conjugate Gradient iterations as soon as the ratio between the actual residual norm and the initial residual norm is less than 99%. The adapted parameter γ_{adapt} is computed by (12) with $\tau = 0.1$ and $\theta = 0.3$ if $|J(\mathbf{q}_{cg}) - J(0.8\mathbf{q}_{cg})| > 1.e-5$ and $\theta = 1.3$ otherwise. Table 1 reports γ_{adapt} and the relative error obtained by `CLSId`(γ_{adapt}) ($PRE(\gamma_{adapt})$) for different noise levels. These values are compared to γ_{opt} and $PRE(\gamma_{opt})$ where the optimal parameter γ_{opt} is computed by applying `CLSId`(γ) to a large set of parameters γ and taking the value γ corresponding to the minimum PRE. The computational performance is reported by the number of steps ℓ and the total number of Gauss Newton iterations $GNit$. Figures 1(a) and 2(a) show the reconstructed and true parameter obtained in the case $\delta = 0.1$ and $\delta = 5.e-3$ respectively. In figure 1(b) the parameter relative errors obtained by choosing the smoothing levels γ_{adapt}, γ_{cg}, $\gamma_0 = \|L\mathbf{q}_0\|_2^2$ are compared to the error obtained by the optimal parameter γ_{opt}. We observe that both $\gamma_{c}g$ and γ_0 give large reconstruction errors while γ_{adapt} allows us to obtain nearly optimal results. Finally the typical behavior of the parameter error obtained by the iterative procedure (12) is shown in figure 2(b).

6. Conclusions

Future work includes the application of different quasi-Newton methods and the extension to non linear time dependent differential models typical of biochemical reactions.

Acknowledgement

The author would like to thank the reviewers for their comments and suggestions.

Test	δ	$PRE(\gamma_{adapt})$	γ_{adapt}	$\ell(GNit)$	γ_{opt}	$PRE(\gamma_{opt})$
T1	1.e-3	9.469e-3	2.423e-3	1 (25)	2.436e-3	9.341e-3
	5.e-3	1.051e-2	2.2424e-3	1 (18)	2.232e-3	1.011e-2
	1.e-2	1.571e-2	2.429e-3	1 (17)	2.916e-3	1.357e-2
T2	1e-03	7.8122e-3	9.0088e-3	6 (63)	9.0284e-3	7.8002e-3
	5e-03	1.5877e-2	9.0306e-3	6 (58)	9.0656e-3	1.5863e-2
	1e-02	1.1775e-2	2.9921e-2	7 (68)	6.9674e-03	2.8706e-02

Table 1. `CLSId` results.

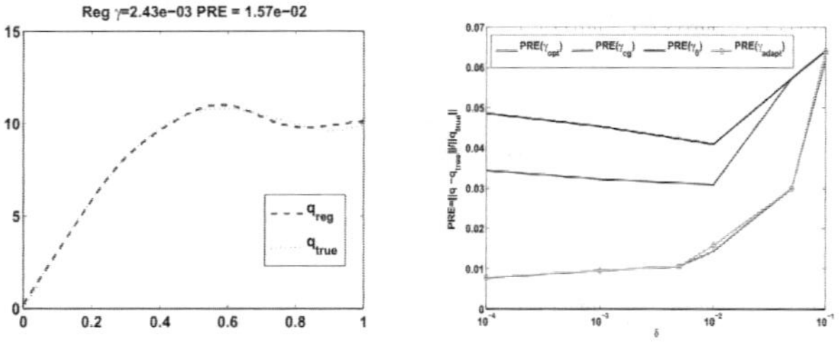

(a) Case $\delta = 1.e-2$ (b) $PRE(\gamma)$ vs δ

Figure 1. Test Problem (13)

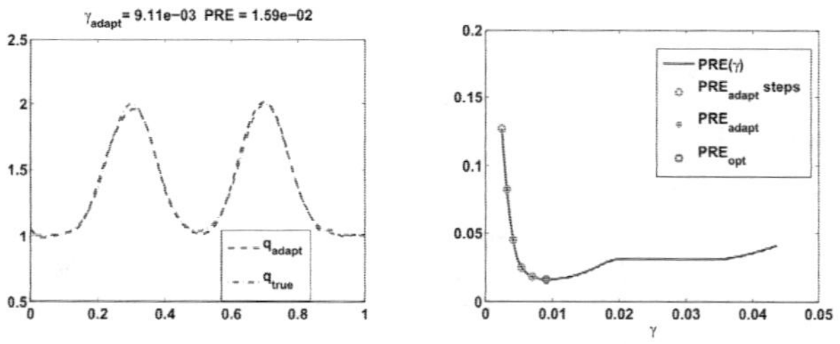

(a) Case $\delta = 5.e-3$ (b) $PRE(\gamma)$ vs γ

Figure 2. Test Problem (14)

References

[1] Becker R, Braack M and Vexler B 2004 Compust. Theory and Modelling **8**

[2] Zama F, Ciavarelli R, Frascari D and Pinelli D 2013 Numerical parameters estimation in models of pollutant transport with chemical reaction *System Modeling and Optimization* (*IFIP Advances in Information and Communication Technology* vol 391) ed Dietmar H and Fredi T (Springer Berlin Heidelberg) pp 547–556 ISBN 978-3-642-36061-9

[3] Ito K and Kunisch K 1990 *SIAM Journal on Control and Optimization* **28** 113–136

[4] Kunisch K and Sachs E W 1992 *SIAM Journal on Numerical Analysis* **29** 1793–1820

[5] B K, A N and O S 2008 *Iterative Regularization Methods for Nonlinear Problems* (de Gruyter Berlin, New York)

[6] B K, a K and B V 2014 *Inverse Problems* **30** 045002

[7] Piccolomini E L and Zama F 2011 *Applied Mathematics and Computation* **217**

[8] Chen K, Piccolomini E and Zama F 2013 *Numerical Algorithms* 1–20

[9] Paige C and Saunders M 1981 *SIAM Journal on Numerical Analysis* **18** 398–405

[10] Åke Björck 1996 *Numerical Methods for Least Squares Problems* (Society for Industrial and Applied Mathematics)

[11] Cartis C, Gould N and Toint P 2009 *BIT*

[12] Jin B and Maass P 2012 *Inverse Problems* **28** 123001

Journal of Physics: Conference Series **657** (2015) 012003 doi:10.1088/1742-6596/657/1/012003

Stochastic level set evolution for nonlinear phase contrast tomography

B Sixou[1]

[1] CREATIS, INSA-Lyon; Inserm, U1044; CNRS UMR 5220; Université Lyon 1; Université de Lyon; 69621, Lyon, France

E-mail: `bruno.sixou@insa-lyon.fr`

Abstract. A stochastic pertubation of a level-set regularization method to find stable inverses for the nonlinear reconstruction of the complex refractive index in in-line phase tomography is presented under the assumption that the index is piecewise constant. The stochastic algorithm is tested on a multi-material object. It is useful to escape the local minima with decrease of the reconstruction errors localized on the boundaries.

1. Introduction

In-line phase tomography is based on a coupling of tomography and phase retrieval. Its aim is to reconstruct the real part of the refractive index [1]. The experimental set-up of Figure 1 shows that the phase contrast can be obtained for several projection angles by letting the X-rays propagate after the object and by recording the Fresnel intensity for one or several propagation distances [2, 3]. Several linear inversion methods have been already studied [4, 5, 6, 7, 8]. The simultaneous nonlinear reconstruction of the real and of the imaginary part of the index has been studied with a Tikhonov regularization [9]. We investigate here a level-set regularization assuming that the ratio of the imaginary to the real part of the complex refractive index is piecewise constant. The nonlinear phase tomography problem is non-convex and the reconstructed solution is a local minimum. We use here stochastic level-set methods to improve the reconstruction obtained with a deterministic level-set scheme. We first summarize the direct problem of the image formation. Then we detail the level-set regularization method and the stochastic approach. The numerical results obtained on noisy simulated data are presented and discussed in the next section before the conclusion.

2. The direct problem

We assume that the real and imaginary parts of the complex refractive index to reconstruct from the Fresnel intensity measurements, denoted as δ and β are defined on a 3D bounded domain (Σ) with spatial coordinates (x, y, z). We denote (x_θ, y_θ, z) be the rotated spatial coordinate system for an angle θ around the z axis (Figure 1). We consider a monochromatic, parallel X-ray beam propagating in the y_θ direction. For a fixed projection angle θ, the interaction can be described by a transmittance function T of the coordinates $X_\theta = (x_\theta, z)$:

$$T_\theta(X_\theta) = \exp[-B_\theta(X_\theta) + i\varphi_\theta(X_\theta)] = a_\theta(X_\theta)\exp[i\varphi_\theta(X_\theta)] \tag{1}$$

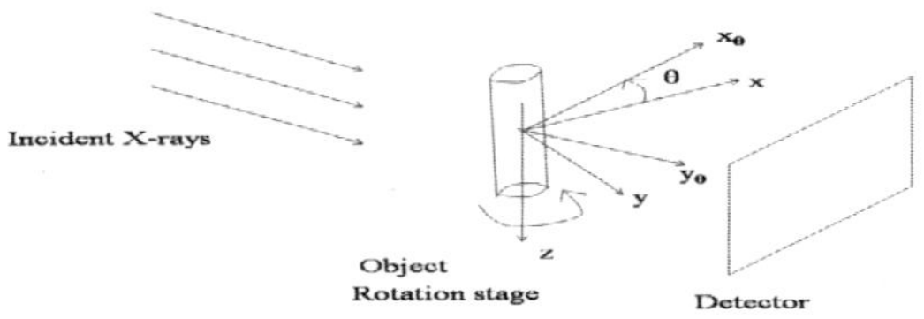

Figure 1. Experimental set-up in propagation based phase contrast tomography with a single propagation-based phase contrast image per projection showing the coordinate system.

.

where $a_\theta(X_\theta)$ is the absorption and $\varphi_\theta(X_\theta)$ the phase shift induced by the object [1]. The phase and the negative logarithm of the absorption, $B_\theta(X) = -\log(a_\theta(X))$, are the projections of the absorption index and refraction index respectively:

$$B_\theta(X_\theta) = \frac{2\pi}{\lambda}\int \beta(y_\theta, X_\theta)dy_\theta \quad \varphi_\theta(X_\theta) = \frac{2\pi}{\lambda}\int (1-\delta(y_\theta, X_\theta))dy_\theta \tag{2}$$

For parallel beam projection, with a beam parallel to the $X = (x, y)$ plane and $f \in L^1(\Sigma)$, the Radon transform is the line integral:

$$Rf(\theta, p, z) = R_\theta f(p, z) = \int_{\vec{t}\in L(\theta,p,z)\cap\Sigma} f(\vec{t})d\vec{t} \tag{3}$$

where $L(\theta, p, z)$ the line at height z defined by $L(\theta, p, z) = \{\tau\bar{\theta}^* + p\bar{\theta} : \tau \in \mathbb{R}\}$, with $\bar{\theta} = (cos(\theta), sin(\theta))$ and $\bar{\theta}^* = (-sin(\theta), cos(\theta))$. With the Radon projection operator, the phase and the absorption can be rewritten $B_\theta(X_\theta) = \frac{2\pi}{\lambda}R_\theta[\beta](X_\theta)$ and $\varphi_\theta(X_\theta) = \frac{2\pi}{\lambda}R_\theta[\delta](X_\theta)$. The Fresnel intensity detected at a distance D after the sample is the 2D convolution, $I_{D,\theta}(X_\theta) = |T_\theta(X_\theta) * P_D(X_\theta)|^2$ where the Fresnel propagator is $P_D(X_\theta) = \frac{1}{i\lambda D}exp(i\frac{\pi}{\lambda D}|X_\theta|^2)$ [1]. The direct nonlinear operator $I_D[\delta, \beta](\theta, X_\theta) = I_{D,\theta}[\delta, \beta](X_\theta)$ is thus the composition of the Fresnel intensity operator and of the Radon projection operator [9].

3. Level-set regularization of in-line phase tomography

3.1. The level-set approach

The phase contrast tomography inverse problem can be regularized with Tikhonov [9] or level-set regularization. Level-set methods have been designed to reconstruct piecewise constant solutions of inverse problems [12, 13]. We assume that δ and β are piecewise constant and that they can take two values δ_1, δ_2 and β_1, β_2 on disjoint measureable subsets Σ_1, Σ_2 such that $\Sigma = \Sigma_1 \cup \Sigma_2$. The unknown functions δ and β are associated with a level-set function of the first order Sobolev space $\eta \in H_1(\Sigma)$, $\beta = \beta_1 + H(\eta)(\beta_2 - \beta_1)$ and $\delta = \delta_1 + H(\eta)(\delta_2 - \delta_1)$. H is the Heaviside distribution, $H(\eta) = 1$ if $\eta > 0$ and 0 otherwise. A variational approach is followed and the following $TV - H_1$ regularization functional is minimized:

$$F[\eta] = \|I_D[\eta] - I_{\delta_n}\|_2^2 + \alpha_1|H[\eta]|_{TV} + \alpha_2\|\eta\|_{H_1}^2 \tag{4}$$

where I_{δ_n} are the noisy intensity data, α_1, α_2 are regularization parameters. To obtain the minimizers, the following smooth approximations of the Heaviside function H has been used, $H_\epsilon(x) = \frac{1+2\epsilon}{2}(erf(x/\epsilon) + 1) - \epsilon$. The length of the boundary between the sets Σ_1 and Σ_2 is penalized with the TV term.

3.2. Implementation of the nonlinear regularization approach

The TV term is dominated by the Sobolev term and can be ignored in a first approximation. For one projection angle θ, the first order optimality condition obtained with the variation of the regularization functional with respect to η can be then written:

$$\delta F[\eta] = I'^{*}_{D,\theta}[\eta][I_{D,\theta}[\eta] - I_{\delta_n}] + \alpha_2(I - \Delta)[\eta] = 0$$

where $I^{*}_{D,\theta}$ denotes the adjoint of the forward operator with respect to L_2 spaces, I the identity and Δ the Laplacian operators. The update is calculated with a Gauss-Newton by $\eta_{k+1} = \eta_k + \delta\eta$ and $\delta\eta$ is the solution of the linear system:

$$([I'^{*}_{D,\theta}[\eta_k]I'_{D,\theta}[\eta_k])\delta\eta + \alpha_2(I - \Delta)\delta\eta = -\delta F[\eta_k]$$

An explicit formula can be derived for the Fréchet derivative of the intensity $I_D[\delta, \beta]$ and its adjoint based on the results obtained for the Tikhonov regularization [9]. Let $T_k = exp(-B[\beta_k] - i\varphi[\delta_k])$, $U_k = [\overline{T_k} * \overline{P_D}]$, $\Delta\beta = \beta_2 - \beta_1$,$\Delta\delta = \delta_2 - \delta_1$, the Fréchet derivative of the operator $I_D[\eta_k]$ at the point η_k is the operator $I'_D[\eta_k] : L^2(\Sigma) \to L^2([0, \pi] \times \mathbb{R}^2)$ defined by:

$$I'_{D,\theta}[\eta_k](\epsilon) = 2Real\{[(-R_\theta H'_\epsilon[\eta_k](\Delta\beta + i\Delta\delta)\epsilon T_k) * P_D]U_k\} \quad (5)$$

where H'_ϵ is the derivative of the smoothed Heaviside function. The adjoint $I'_D[\eta_k]^* : L^2([0, \pi] \times \mathbb{R}^2) \to L^2(\Sigma)$ is defined by $I'_{D,\theta}[\eta_k]^*(u) = 2Real(v_1 + v_2)$ with:

$$v_1 = (\beta_2 - \beta_1)H'_\epsilon[\eta_k]R^*_\theta\{[-u(T_k * P_D) * \overline{P_D}]\overline{T_k}\} \quad (6)$$

$$v_2 = (\delta_2 - \delta_1)H'_\epsilon[\eta_k]R^*_\theta\{[u(T_k * P_D) * \overline{P_D}]i\overline{T_k}\} \quad (7)$$

3.3. Stochastic level-set evolution

The deterministic level-set minimization algorithm may be trapped in local minima. The aim of this work is to obtain a stochastic evolution of the boundary curve to improve the reconstruction with a new stochastic level-set method. Level-set methods have been much studied for image processing tasks [14] but the stochastic level-set approach has been proposed only for segmentation [10] or binary tomography [11] . Let $(\Omega, \mathcal{F}, P)$ be a probability space, in order to obtain the global minimum of a function $g : \mathbb{R}^m \to \mathbb{R}$, a random trajectory $X(t)$ governed by a diffusion process of the type $dX(t) = -\nabla g(X(t))dt + \rho(t)dW(t)$ is often used [15, 16, 17, 18], where $W = (W_1(t), ..., W_m(t))$ is the standard m-dimensional Brownian motion and $\rho(t)$ the noise strength. Under appropriate condition on the annealing schedule $\rho(t)$ and the function g, X(t) converges towards the global minimum of g. The same formalism can be applied to the level-set function. The evolution of the boundary curve is independent of the level-set function used for its representation with the Stratanovich integral formalism [10]. We thus propose to improve the reconstruction image obtained with the deterministic level-set evolution with the following stochastic partial differential equation for the level-set function η, given for $\vec{r} \in \Sigma$ by:

$$d\eta(\vec{r}, t) = \delta\eta(\vec{r}, t) + \rho(t)|\nabla\eta(\vec{r}, t)|odW(t) \quad (8)$$

where o denotes the Stratanovitch convention [19] and $\delta\eta$ is the deterministic change calculated as explained in Section 3.2. We obtain the following Itô stochastic differential equation with the definition of the Stratanovich integral [10, 19]:

$$d\eta(\vec{r}, t) = \delta\eta + \rho(t)|\nabla\eta(\vec{r}, t)|dW(t) + \frac{1}{2}\rho(t)(\Delta\eta(\vec{r}, t) - |\nabla\eta(\vec{r}, t)|div(\frac{\nabla\eta(\vec{r}, t)}{|\nabla\eta(\vec{r}, t)|}) \quad (9)$$

In this study, the deterministic level-set and stochastic level-set schemes are applied successively on random time intervals with an intermittent diffusion [17]. For the stochastic evolution, the time interval lengths and the diffusion strengths μ are chosen at random in the range $[0, T_{max}]$ and $[0, \mu_{max}]$ where $\mu_{\max}$ is the scale for the diffusion strength and T_{max} is the scale for the diffusion time.

4. Results and Discussion

4.1. Simulation details

The deterministic and intermittent stochastic level-set algorithms for the refraction index recovery are compared on an multi-material object, with an Al homogeneous cylinder of 20 μm in diameter and 110 μm in height embedded in PMMA. Let $\mu = \frac{4\pi\beta}{\lambda}$, the δ and μ values used for PMMA and Al for 24 keV X-rays are 4.62810 10^{-7}, 9.396 10^{-7} and 41.2 m^{-1}, 502.6 m^{-1} respectively. The β and δ values were discretized on a regular grid with a pixel size of 1.5 μm. The cylinder is included in a rectangular volume of size $N_1 \times N_1 \times N_2$ with $N_1 = 74$ and $N_2 = 109$. Horizontal sections of the original maps to be retrieved δ^* and β^* are displayed in Fig.2.

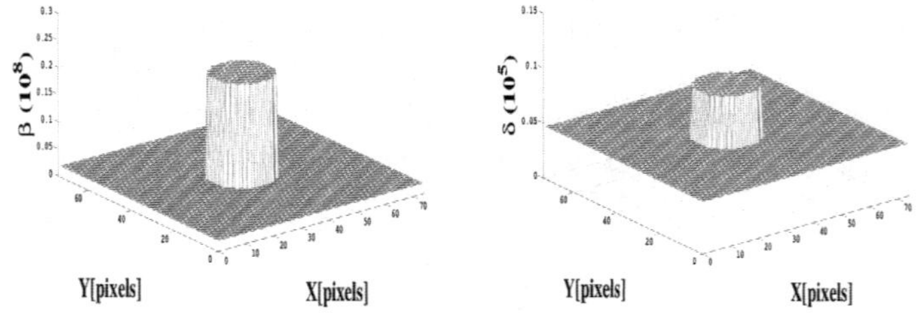

Figure 2. True beta and delta map.

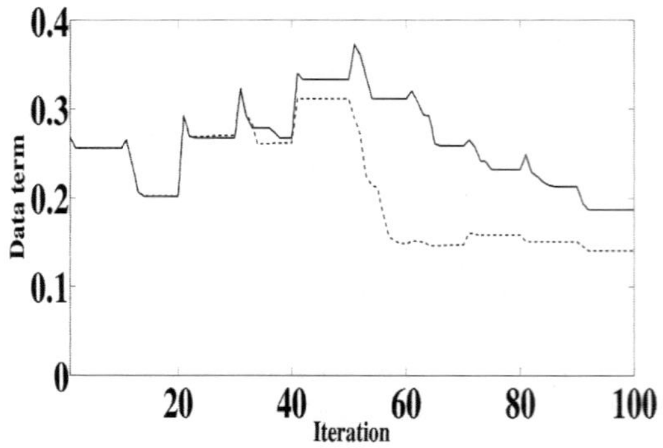

Figure 3. Evolution of the data term $\|I_{D,\theta}[\eta] - I_{\delta_n}\|$ with the iterations for the deterministic (plain line) and intermittent stochastic algorithms (dotted line).

The discrete approximation of the Radon transform is the projection operator implemented in the Mablab Toolbox. The reconstruction was performed with 400 projection angles, the X-ray wavelength $\lambda = 0.5166$ $\mathring{A}$ (24keV) and the distance $D = 100$ mm. The intensity data were corrupted with additive Gaussian white noise with a peak-to-peak signal to noise ratio (PPSNR) of 18 dB. To obtain a good accuracy, the ϵ parameter was fixed to $\epsilon = 0.03$. The initial level-set function chosen is $\theta_0 = 0$. The regularization parameters were chosen to obtain

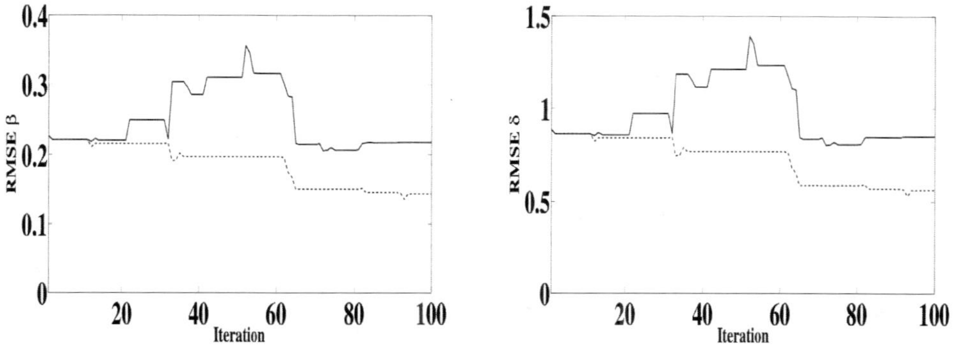

Figure 4. Evolution of the normalized mean square error for β and δ with the iterations for the deterministic (plain line) and stochastic algorithms (dotted line).

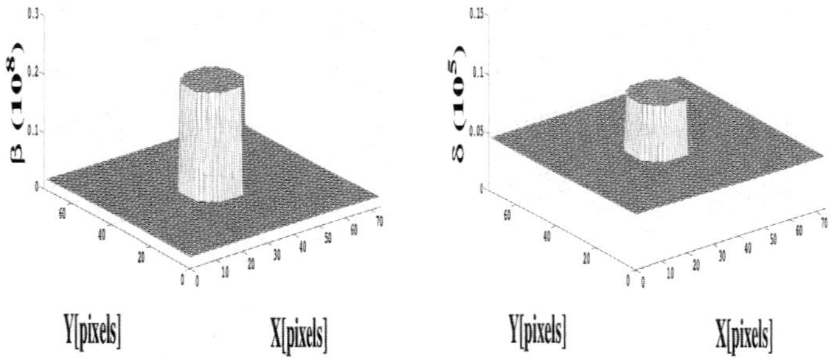

Figure 5. Reconstructed β and δ maps.

the best decrease of the data term. A first reconstruction is performed with the deterministic algorithm. Then the intermittent diffusion is applied with projection angles chosen at random. For a fixed projection angle, the deterministic and stochastic steps are thus applied successively. The numbers of iterations are chosen randomly with a uniform distribution in $[1, 50]$ and the noise strength in the range $[1, 10^{-3}]$. For the simulation of Eq.(9), we use an explicit scheme with finite differences, the WENO scheme [20] with $\Delta x = 0.1$ and $\Delta t = 0.01$. The level-set function is periodically reinitialized. For comparison, the deterministic level-set algorithm is applied iteratively with projection angles also chosen at random and periodic reinitilizations of η.

4.2. Results

The evolutions of the the relative data term $\|I_{D,\theta}[\eta] - I_{\delta_n}\|/\|I_{\delta_n}\|$ are displayed on Figure 3 for the deterministic and intermittent stochastic algorithms starting from the initial reconstruction. The deterministic minimization can not escape the local minimum corresponding to the initial reconstructed δ and β volumes. A larger decrease is obtained for the deviation between the discrepancy term and the noise level with the stochastic scheme. In order to have a more quantitative information about the efficiency of the method, the evolution of the normalized mean square error using the $L_2(\Sigma)$ norm, $\|\delta^* - \delta\|_2/\|\delta^*\|_2$ and $\|\beta^* - \beta\|_2/\|\beta^*\|_2$, are displayed as a function of the number of iterations for the deterministic and stochastic algorithms on Figure 4. The deterministic algorithm is not efficient to achieve lower reconstruction errors.

On the contrary, the reconstruction errors for β and δ are much reduced with the intermittent stochastic diffusion. Figure 5 displays some horizontal section of the real and imaginary part of the reconstructed index map obtained for the minimum of the discrepancy term. These figures show that the reconstruction errors have been significantly reduced.

5. Conclusion

In this work, we have investigated the nonlinear inverse problem associated to the reconstruction of the real and complex part of the refractive index in phase contrast tomography. The a priori information that the real and imaginary parts δ and β are piecewise constant is introduced and the regularization is performed on the level-set function. The main contribution of this work is to compare the reconstruction results obtained for a deterministic and stochastic evolution of the level-set function on an homogeneous cylinder phantom. The stochastic algorithm is useful to escape the local minimum obtained with the level-set regularization and it leads to a larger decrease of the reconstruction errors localized on the boundaries.

References

[1] M.Langer, P.Cloetens, A.Pacureanu and F.Peyrin "X-ray in-line phase tomography of multimaterial objects", *Optics Letters*, vol.37, pp.2151-2154 , 2012.
[2] M.Langer, P.Cloetens, J.P.Guigay and F.Peyrin "Quantitative comparison of direct phase retrieval algorithms in in-line phase tomography", *Medical Physics*, vol.35, pp.4556-4565 , 2008.
[3] K.A.Nugent "Coherent methods in the X-rays science", *Advances in Physics*, vol.59, pp.1-99 , 2010.
[4] D.Paganin, S.C.Mayo, T.E.Gureyev, P.R.Miller and S.W.Wilkins "Simulateneous phase and amplitude extraction from a single defocused image of a homogeneous object", *J.Microsc.*, vol.206, pp.33-40 , 2002.
[5] D.Paganin, "Coherent X-ray optics", *Oxford University Press, New York* , 2006.
[6] T.E.Gureyev, D.M.Paganin, G.R.Myers, Y.I.Nesterest and S.W.Wilkins "Phase-and-amplitude computer tomography", *Applied Physics Letters*, vol.89, pp.034102-034104 , 2006.
[7] J.P.Guigay , M.Langer , R.Boistel and P.Cloetens "A mixed contrast transfer and transport of intensity approach for phase retrieval in the Fresnel region", *Opt.Lett.*, vol.32, pp.1617-1629 , 2007.
[8] M.A.Beltran, D.M.Paganin, K.Uesugi and M.J.Kitchen "2D and 3D X-ray phase retrieval of multi-material objects using a single defocus distance", *Optics Express*, vol.18, pp.6423-6436, 2010.
[9] B.Sixou "Reconstruction of the complex refractive index in nonlinear phase contrast tomography", *Inverse Problems in Science and Engineering*, 2014.
[10] O.Juan, R.Keriven and G.Postelnicu "Stochastic Motion and the Level-Set Method in Computer Vision: Stochastic Active Contours", *International Journal of Computer Vision*, vol.69, pp.7-25 , 2006.
[11] B.Sixou "Binary tomography reconstruction with stochastic level-set methods", *IEEE Signal Processing Letters*, vol.22, pp.920-924 , 2015.
[12] A.Egger and L.Leitao "Nonlinear regularization for ill-posed problems with piecewise constant or strongly varying solutions", *Inverse Problems*, vol.25, pp.115014 , 2009.
[13] A.DeCezaro, A.Leitao and X.C.Tai "On multiple level-set regularization methods for inverse problems", *Inverse Problems*, vol.25, pp.035004, 2009.
[14] G.Aubert and P.Kornprobst, "Mathematical Problems in Image Processing, Partial Differential Equations and the Calculus of Variations", *Applied Mathematical Sciences* , 147, Springer 2006.
[15] S.Geman and C.R.Hwang, "Diffusion for global optimization", *J.Control Optim.*, vol.24, pp.1031-1043, 1986.
[16] P.Parpas and B.Rustem, "Convergence analysis a a global optimization algorithm using stochastic differential equations", *J.Control Optim.*, vol.45, pp.95-110, 2009.
[17] S.Chow, T.Yang and H.Zhou,, "Global optimization by intermittent diffusion", *J.Control Optim.*, vol.45, pp.95-110, 2009.
[18] T.S.Chiang, C.R.Hwang and S.J.Sheu, "Diffusion for global optimization in $\mathbb{R}^n$", *SIAM J.Control Optim.*, vol.25, pp.737-753, 1987.
[19] G.Da Prato and J.Zabczyk, "Stochastic equations in infinite dimensions", *Encyclopedia of Mathematics and Its Applications, 44. Cambridge:Cambridge University Press* , 1992.
[20] G.S.Jiang and D.Peng, "Weigthed ENO schemes for Hamilton-Jacobi equations", *SIAM J.Control Optim.*, vol.21, pp.2126-2143, 2000.

An *a posteriori* error estimator for shape optimization: application to EIT

M Giacomini[1,2] , O Pantz[1] and K Trabelsi[2]

[1] CMAP Ecole Polytechnique, Route de Saclay, 91128 Palaiseau, France (M. Giacomini and O. Pantz are members of the DEFI project at INRIA Saclay Ile-de-France)

[2] DRI Institut Polytechnique des Sciences Avancées, 7-9 rue M. Grandcoing, 94200 Ivry-sur-Seine, France

E-mail: matteo.giacomini@polytechnique.edu

Abstract. In this paper we account for the numerical error introduced by the Finite Element approximation of the shape gradient to construct a guaranteed shape optimization method. We present a goal-oriented strategy inspired by the complementary energy principle to construct a constant-free, fully-computable *a posteriori* error estimator and to derive a certified upper bound of the error in the shape gradient. The resulting Adaptive Boundary Variation Algorithm (ABVA) is able to identify a genuine descent direction at each iteration and features a reliable stopping criterion for the optimization loop. Some preliminary numerical results for the inverse identification problem of Electrical Impedance Tomography are presented.

1. Introduction

A classical approach to shape optimization and shape identification problems relies on the application of the Steepest Descent Method using the so-called shape gradient. In most applications, the shape gradient depends on the solution of a PDE which can only be solved approximately. Thus stopping criteria based on its norm are never fulfilled if the *a priori* given tolerance is too small. In this paper, we evaluate the discretization error in the shape gradient relying on the well-established results on goal-oriented error estimators [1] and we propose a guaranteed shape optimization strategy. Several works in the literature have focused on Adaptive Finite Element Method for shape optimization problems ([2], [3], [4], [5]) but to the best of our knowledge, all the proposed approaches used error estimators only to drive mesh adaptivity without accounting for the error in the shape gradient itself.

After briefly recalling the framework of shape optimization and shape identification problems in section 2 and the notion of shape gradient in section 3, we introduce the Adaptive Boundary Variation Algorithm for shape optimization problems (Section 4). In section 5 we present some preliminary numerical simulations for the inverse problem of Electrical Impedance Tomography and a summary of our results ends the paper.

2. Shape optimization and shape identification problems

Let us consider an open domain $\Omega \subset \mathbb{R}^d$ $(d \geq 2)$ with Lipschitz boundary $\partial\Omega$. Let V_Ω be a separable Hilbert space depending on Ω, we define $u_\Omega \in V_\Omega$ to be the solution of a state equation

$$a_\Omega(u_\Omega, \delta u) = F_\Omega(\delta u) \quad \forall \delta u \in V_\Omega \tag{1}$$

where $a_\Omega(\cdot,\cdot)$ is a continuous and coercive bilinear form on V_Ω and $F_\Omega(\cdot)$ is a continuous linear form on V_Ω, both of them depending on Ω. Under these assumptions, problem (1) has a unique solution.
We introduce a cost functional $J(\Omega) = j(\Omega, u_\Omega)$ which depends on the domain Ω itself and on the solution u_Ω of the state equation. We consider the following shape optimization problem

$$\min_{\Omega \in \mathcal{U}_{\text{ad}}} J(\Omega) \tag{2}$$

where $\mathcal{U}_{\text{ad}}$ is the set of admissible domains in $\mathbb{R}^d$. Within this framework, problem (2) may be viewed as a PDE-constrained optimization problem where we aim at minimizing $j(\Omega, u)$ under the constraint $u = u_\Omega$.

The case of Electrical Impedance Tomography
Let $\mathcal{D}$ be a domain featuring an inclusion Ω. We define the conductivity k_Ω as a piecewise constant function equal to k_I inside the inclusion Ω and k_E otherwise ($k_I, k_E > 0$). We are interested in identifying the shape and the location of the inclusion, fitting given boundary measurements g and U_D of the flux and the potential. Thus we introduce $u_{\Omega,N}, u_{\Omega,D} \in H^1(\mathcal{D})$ such that $u_{\Omega,D} = U_D$ on $\partial\mathcal{D}$, solutions of respectively a Neumann and a Dirichlet Boundary Value Problem:

$$\int_{\mathcal{D}} \left(k_\Omega \nabla u_{\Omega,i} \cdot \nabla \delta u_i + u_{\Omega,i} \delta u_i \right) dx = \begin{cases} \int_{\partial \mathcal{D}} g \delta u_i ds & \forall \delta u_i \in H^1(\mathcal{D}) \ , \ i = N \\ 0 & \forall \delta u_i \in H^1_0(\mathcal{D}) \ , \ i = D \end{cases} \tag{3}$$

The inverse identification problem of Electrical Impedance Tomography may be written as in (2) and we seek the open subset that minimizes the Kohn-Vogelius functional [6]:

$$J(\Omega) = \frac{1}{2} \int_{\mathcal{D}} \left(k_\Omega \left| \nabla (u_{\Omega,N} - u_{\Omega,D}) \right|^2 + |u_{\Omega,N} - u_{\Omega,D}|^2 \right) dx. \tag{4}$$

3. Differentiation with respect to the shape

Let X be a Banach space and $\theta \in X$ be an admissible smooth deformation of Ω. The objective functional J is said to be X-differentiable at $\Omega \in \mathcal{U}_{ad}$ if there exists a continuous linear form $dJ(\Omega)$ on X such that $\forall \theta \in X$ we have $J((\text{Id} + \theta)\Omega) = J(\Omega) + \langle dJ(\Omega), \theta \rangle + o(\theta)$.
Let us introduce the Lagrangian functional, defined for every admissible open set Ω and every $u, p \in V_\Omega$ by

$$\mathcal{L}(\Omega, u, p) = j(\Omega, u) + a_\Omega(u, p) - F_\Omega(p), \tag{5}$$

and the adjoint state $p_\Omega \in V_\Omega$ such that $\forall \delta p \in V_\Omega$

$$a^*_\Omega(p_\Omega, \delta p) + \left\langle \frac{\partial j}{\partial u}(\Omega, u_\Omega), \delta p \right\rangle = 0 \tag{6}$$

Let us now introduce a diffeomorphism $\varphi : \mathbb{R}^d \to \mathbb{R}^d$. For every admissible set Ω, and for all $u, p \in V_\Omega$, we set $\Omega_\varphi := \varphi(\Omega)$, $u_\varphi := u \circ \varphi^{-1}$ and $p_\varphi := p \circ \varphi^{-1}$. We admit that $u \mapsto u_\varphi$ is a one-to-one map between V_Ω and $V_{\varphi(\Omega)}$. We define the material derivative $\frac{\partial \mathcal{L}}{\partial \varphi}$ of $\mathcal{L}$ as the derivative of the map $\varphi \mapsto \mathcal{L}(\Omega_\varphi, u_\varphi, p_\varphi)$ evaluated in $\varphi = \text{Id}$. Hence, from the fast derivation method of Céa [7] the shape gradient reads as

$$\langle dJ(\Omega), \theta \rangle = \left\langle \frac{\partial \mathcal{L}}{\partial \varphi}(\Omega, u_\Omega, p_\Omega), \theta \right\rangle. \tag{7}$$

The case of Electrical Impedance Tomography
In order to compute the shape gradient of (4), we need to determine the adjoint solutions $p_{\Omega,N}$ and $p_{\Omega,D}$ associated with the states $u_{\Omega,N}$ and $u_{\Omega,D}$: the Kohn-Vogelius problem is self-adjoint and we get that $p_{\Omega,N} = u_{\Omega,N} - u_{\Omega,D}$ and $p_{\Omega,D} = 0$. The most common approach in the literature is based on an Eulerian viewpoint and leads to a surface expression of the shape gradient

$$\langle dJ(\Omega),\theta\rangle = \frac{1}{2}\int_{\partial\Omega}\left([\![k_\Omega]\!]\left(\left|\frac{\partial u_{\Omega,N}}{\partial\tau}\right|^2 - \left|\frac{\partial u_{\Omega,D}}{\partial\tau}\right|^2\right) - [\![k_\Omega^{-1}]\!]\left(\left|k_\Omega\frac{\partial u_{\Omega,N}}{\partial n}\right|^2 - \left|k_\Omega\frac{\partial u_{\Omega,D}}{\partial n}\right|^2\right)\right)(\theta\cdot n)ds \tag{8}$$

where n is the outward normal to $\partial\Omega$, τ is the tangential direction to $\partial\Omega$ and $[\![k_\Omega]\!] = k_E - k_I$ and $[\![k_\Omega^{-1}]\!] = k_E^{-1} - k_I^{-1}$ are the jumps across $\partial\Omega$. Let us now introduce the following operator:

$$\langle G(\Omega,u),\theta\rangle = \frac{1}{2}\int_{\mathcal{D}}\left(k_\Omega M(\theta)\nabla u\cdot\nabla u - \nabla\cdot\theta\ u^2\right)dx \tag{9}$$

where $M(\theta) = \nabla\theta + \nabla\theta^T - (\nabla\cdot\theta)\,\mathrm{Id}$. From (7), we get a volumetric expression of the shape gradient

$$\langle dJ(\Omega),\theta\rangle = \langle G(\Omega,u_{\Omega,N}) - G(\Omega,u_{\Omega,D}),\theta\rangle \tag{10}$$

We refer to [8] for more details on the differentiation of the Kohn-Vogelius functional and its application to the identification of discontinuities of the conductivity parameter.

4. The Adaptive Boundary Variation Algorithm for shape optimization

In algorithm 1, we present the Boundary Variation Algorithm, that is the application of the classical Steepest Descent Method to a shape optimization problem. After computing the solution of the state equation, we derive the expression of the shape gradient and we compute the corresponding descent direction to deform the computational domain.

Algorithm 1: Continuous gradient method - The Boundary Variation Algorithm

Given the domain Ω_0, set $\ell = 0$ and iterate:
1. Compute the solutions u_{Ω_ℓ} and p_{Ω_ℓ} of the state and adjoint equations;
2. Compute a descent direction $\theta_\ell \in X$ solving $(\theta_\ell,\delta\theta)_X + \langle dJ(\Omega_\ell),\delta\theta\rangle = 0 \quad \forall\delta\theta\in X$;
3. Identify an admissible step μ_ℓ and update the domain $\Omega_{\ell+1} = (\mathrm{Id} + \mu_\ell\theta_\ell)\Omega_\ell$;
4. While $|\langle dJ(\Omega_\ell),\theta_\ell\rangle| > \mathrm{tol}$, $\ell = \ell+1$ and repeat.

This method relies on the computation at each iteration of a direction θ such that the shape gradient of the objective functional in this direction is strictly negative, that is we seek θ such that $\langle dJ(\Omega),\theta\rangle < 0$.

Let us now introduce the discretized solutions u_Ω^h and p_Ω^h of the state and adjoint equations using the Finite Element Method. We consider for both u_Ω^h and p_Ω^h the approximation in the space of linear Lagrangian Finite Element functions. In the same fashion, we define the Finite Element discretization of the shape gradient as

$$\langle d_hJ(\Omega),\theta\rangle = \left\langle\frac{\partial\mathcal{L}}{\partial\varphi}(\Omega,u_\Omega^h,p_\Omega^h),\theta\right\rangle \tag{11}$$

The discretized descent direction $\theta^h \in X$ is computed by solving problem (12) using linear Lagrangian Finite Elements.

$$(\theta^h,\delta\theta^h)_X + \langle d_hJ(\Omega),\delta\theta^h\rangle = 0 \quad \forall\delta\theta^h\in X \tag{12}$$

We remark that due to the numerical approximation, it is not sufficient that $\langle d_hJ(\Omega),\theta^h\rangle < 0$ for $J(\Omega)$ to decrease along the direction θ^h. Moreover, if the approximation error is too large the stopping criterion (Algorithm 1 - step 4) will usually not be fulfilled. In this paper we present a strategy to bypass this issue by providing a computable upper bound of the error E^h in the approximation of the shape gradient and by seeking a certified direction θ^h such that $\langle d_hJ(\Omega),\theta^h\rangle + |E^h| < 0$.

4.1. Numerical error in the shape gradient

Let us write the numerical error E^h in the shape gradient as follows

$$\begin{aligned} E^h = \langle dJ(\Omega) - d_h J(\Omega), \theta^h \rangle &= \left\langle \frac{\partial \mathcal{L}}{\partial \varphi}(\Omega, u_\Omega, p_\Omega) - \frac{\partial \mathcal{L}}{\partial \varphi}(\Omega, u_\Omega^h, p_\Omega^h), \theta^h \right\rangle \\ &\simeq \frac{\partial^2 \mathcal{L}}{\partial \varphi \partial u}(\Omega, u_\Omega^h, p_\Omega^h)[\theta^h, u_\Omega - u_\Omega^h] + \frac{\partial^2 \mathcal{L}}{\partial \varphi \partial p}(\Omega, u_\Omega^h, p_\Omega^h)[\theta^h, p_\Omega - p_\Omega^h] \end{aligned} \tag{13}$$

In order to compute an upper bound of the error (13), we use the well-established framework of *a posteriori* error estimators for a Quantity of Interest [1]. For this purpose we introduce two adjoint problems, each of which is associated with one term on the right-hand side of (13) and we seek $r_\Omega, s_\Omega \in V_\Omega$ such that $\forall \delta r, \delta s \in V_\Omega$

$$a_\Omega^*(r_\Omega, \delta r) = \frac{\partial^2 \mathcal{L}}{\partial \varphi \partial u}(\Omega, u_\Omega^h, p_\Omega^h)[\theta^h, \delta r] \quad , \quad a_\Omega^*(s_\Omega, \delta s) = \frac{\partial^2 \mathcal{L}}{\partial \varphi \partial p}(\Omega, u_\Omega^h, p_\Omega^h)[\theta^h, \delta s] \tag{14}$$

By plugging (14) into (13) and exploiting Galerkin orthogonality and Cauchy-Schwarz inequality, we may derive the upper bound $\overline{E}$ for the numerical error in the shape gradient:

$$|E^h| \le \overline{E} := \|r_\Omega - r_\Omega^h\|_\Omega \, \|u_\Omega - u_\Omega^h\|_\Omega + \|s_\Omega - s_\Omega^h\|_\Omega \, \|p_\Omega - p_\Omega^h\|_\Omega \tag{15}$$

where $\|\cdot\|_\Omega$ is the energy-norm induced by the bilinear form $a_\Omega(\cdot,\cdot)$.
We remark that the criterion $\langle d_h J(\Omega), \theta^h \rangle + \overline{E} < 0$ ensures that θ^h is a certified genuine descent direction, whether it is the solution of equation (12) or not. In particular, this also stands when the latter problem is only solved approximately.

4.2. Certified Steepest Descent Method

We may now present the resulting Adaptive Boundary Variation Algorithm arising from the coupling of the classical Boundary Variation Algorithm for shape optimization with the goal-oriented estimator of the error in the shape gradient.

Algorithm 2: Discretized gradient method - The Adaptive Boundary Variation Algorithm

Given the domain Ω_0, set $\ell = 0$ and iterate:
1. Construct a coarse mesh for Ω_ℓ;
2. Compute the solutions $u_{\Omega_\ell}^h$ and $p_{\Omega_\ell}^h$ of the state and adjoint equations;
3. Compute a descent direction $\theta_\ell^h \in X$ solving $(\theta_\ell^h, \delta\theta^h)_X + \langle d_h J(\Omega_\ell), \delta\theta^h \rangle = 0 \quad \forall \delta\theta^h \in X$;
4. Compute an upper bound $\overline{E}$ of the numerical error $|E^h|$:
   (a) Compute the solutions $r_{\Omega_\ell}^h$ and $s_{\Omega_\ell}^h$ to estimate the error in the QoI;
   (b) Compute $\overline{E} = \|r_{\Omega_\ell} - r_{\Omega_\ell}^h\|_\Omega \, \|u_{\Omega_\ell} - u_{\Omega_\ell}^h\|_\Omega + \|s_{\Omega_\ell} - s_{\Omega_\ell}^h\|_\Omega \, \|p_{\Omega_\ell} - p_{\Omega_\ell}^h\|_\Omega$;
5. If $\langle d_h J(\Omega_\ell), \theta_\ell^h \rangle + \overline{E} \ge 0$, refine the mesh and go to 2;
6. Identify an admissible step μ_ℓ and update the domain $\Omega_{\ell+1} = (\mathrm{Id} + \mu_\ell \theta_\ell^h)\Omega_\ell$;
7. While $|\langle d_h J(\Omega_\ell), \theta_\ell^h \rangle| + \overline{E} > \mathrm{tol}$, $\ell = \ell + 1$ and repeat.

The case of Electrical Impedance Tomography

Since the Kohn-Vogelius problem is self-adjoint, (15) reduces to

$$\overline{E} = \|r_{\Omega,N} - r_{\Omega,N}^h\|_\Omega \, \|u_{\Omega,N} - u_{\Omega,N}^h\|_\Omega + \|r_{\Omega,D} - r_{\Omega,D}^h\|_\Omega \, \|u_{\Omega,D} - u_{\Omega,D}^h\|_\Omega \tag{16}$$

where $r_{\Omega,N} \in H^1(\mathcal{D})$ and $r_{\Omega,D} \in H_0^1(\mathcal{D})$ are the solutions of the following adjoint problems introduced to construct the estimator in the Quantity of Interest:

$$\int_{\mathcal{D}} \Big(k_\Omega \nabla r_{\Omega,i} \cdot \nabla \delta r_i + r_{\Omega,i} \delta r_i \Big) dx = \frac{\partial G}{\partial u}(\Omega, u_{\Omega,i}^h)[\theta^h, \delta r_i] \quad \begin{cases} \forall \delta r_i \in H^1(\mathcal{D}) \ , \ i = N \\ \forall \delta r_i \in H_0^1(\mathcal{D}) \ , \ i = D \end{cases} \tag{17}$$

To fully compute the estimate (16), we use the complementary energy principle. For the state equations, we introduce the dual flux variables $\sigma_{\Omega,N}, \sigma_{\Omega,D} \in H(\text{div}) = \{\sigma \in L^2(\mathcal{D};\mathbb{R}^d) : \text{div}\, \sigma \in L^2(\mathcal{D})\}$ such that $\sigma_{\Omega,N} \cdot n = g$ on $\partial\mathcal{D}$ which satisfy $\forall \delta\sigma_N, \delta\sigma_D \in H(\text{div})$ such that $\delta\sigma_N \cdot n = 0$ on $\partial\mathcal{D}$

$$\int_{\mathcal{D}} \Big(k_\Omega^{-1} \sigma_{\Omega,i} \cdot \delta\sigma_i + (\nabla \cdot \sigma_{\Omega,i})(\nabla \cdot \delta\sigma_i) \Big) dx = \begin{cases} 0 & \text{for } i = N \\ \int_{\partial\mathcal{D}} U_D (\delta\sigma_i \cdot n) ds & \text{for } i = D \end{cases} \tag{18}$$

Let $\sigma_{\Omega,N}^h$ and $\sigma_{\Omega,D}^h$ be the dual fluxes discretized using the lowest order Raviart-Thomas Finite Elements. We obtain the following bound for the energy-norm of the error in the state equations:

$$|\!|\!| u_{\Omega,i} - u_{\Omega,i}^h |\!|\!|_\Omega^2 \le \int_{\mathcal{D}} \Big(k_\Omega^{-1} |\sigma_{\Omega,i}^h - k_\Omega \nabla u_{\Omega,i}^h|^2 + |\nabla \cdot \sigma_{\Omega,i}^h - u_{\Omega,i}^h|^2 \Big) dx \tag{19}$$

In a similar fashion, we introduce the dual fluxes $\xi_{\Omega,N}, \xi_{\Omega,D} \in H(\text{div})$ such that $\xi_{\Omega,N} \cdot n = 0$ on $\partial\mathcal{D}$ satisfying $\forall \delta\xi_N, \delta\xi_D \in H(\text{div})$ such that $\delta\xi_N \cdot n = 0$

$$\int_{\mathcal{D}} \Big(k_\Omega^{-1} \xi_{\Omega,i} \cdot \delta\xi_i + (\nabla \cdot \xi_{\Omega,i})(\nabla \cdot \delta\xi_i) \Big) dx = \int_{\mathcal{D}} \Big((\nabla \cdot \theta^h) u_{\Omega,i}^h \nabla \cdot \delta\xi_i - M(\theta^h) \nabla u_{\Omega,i}^h \cdot \delta\xi_i \Big) dx \tag{20}$$

and via their RT_0 approximations $\xi_{\Omega,i}^h$'s, we derive the upper bound of the adjoint errors

$$|\!|\!| r_{\Omega,i} - r_{\Omega,i}^h |\!|\!|_\Omega^2 \le \int_{\mathcal{D}} \Big(k_\Omega^{-1} |\xi_{\Omega,i}^h + k_\Omega M(\theta^h) \nabla u_{\Omega,i}^h - k_\Omega \nabla r_{\Omega,i}^h|^2 + |\nabla \cdot \xi_{\Omega,i}^h - (\nabla \cdot \theta^h) u_i^h - r_{\Omega,i}^h|^2 \Big) dx \tag{21}$$

Eventually, by plugging (19) and (21) into (16), we are able to explicitly compute the bound $\overline{E}$.

5. Numerical results

We apply the ABVA to the identification of an unknown inclusion Ω inside a circular domain $\mathcal{D}$ using one boundary measurement. The deformation of the inclusion (Algorithm 2 - step 6) is performed by moving the whole computational mesh. In figure 1c, we present the reconstructed interface after 25 iterations of the procedure; figure 1d confirms that the ABVA identifies a genuine descent direction at each iteration, generating a sequence of minimizing shapes such that the objective functional is monotonically decreasing and provides a guaranteed stopping criterion. Moreover, in figure 1e we observe that the number of Degrees of Freedom remains small until the reconstructed interface approaches the real inclusion thus coarse meshes prove to be reliable during the initial iterations of the algorithm.

In figure 2, we report the results of a more complex case in which two inclusions are sought. We use ten boundary measurements and the ABVA is still able to identify a genuine descent direction at each iteration (Fig. 2b) but the reconstructed interface appears to be less precise (Fig. 2a). This phenomenon depends on the severe ill-posedness of the problem and the diffusive nature of the state equation is responsible for the loss of the information in the middle of the domain, even when using several measurements. This results in the ability of the algorithm to reconstruct the inclusions near the boundary of the domain whereas the inner interfaces appear more difficult to identify and with a less precise outcome. As previously observed, figure 2c confirms that the ABVA allows to use the same meshes for several iterations increasing the number of DOFs only when the descent direction is no more validated.

6. Conclusion

Introducing an *a posteriori* error estimator for the shape gradient provides quantitative information to define a reliable stopping criterion for the algorithm. Nevertheless, solving the

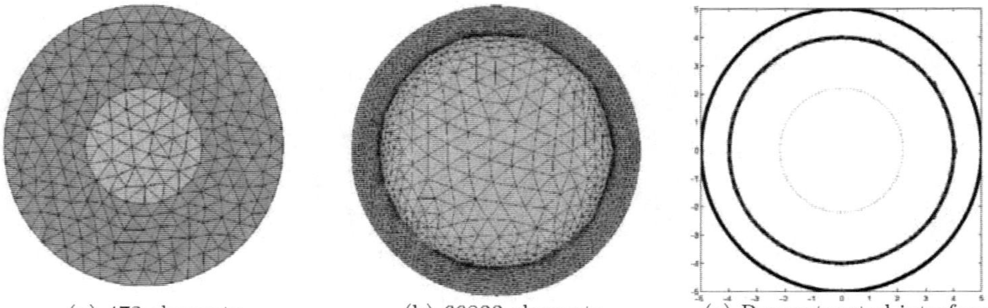

(a) 472 elements. (b) 66833 elements. (c) Reconstructed interface.

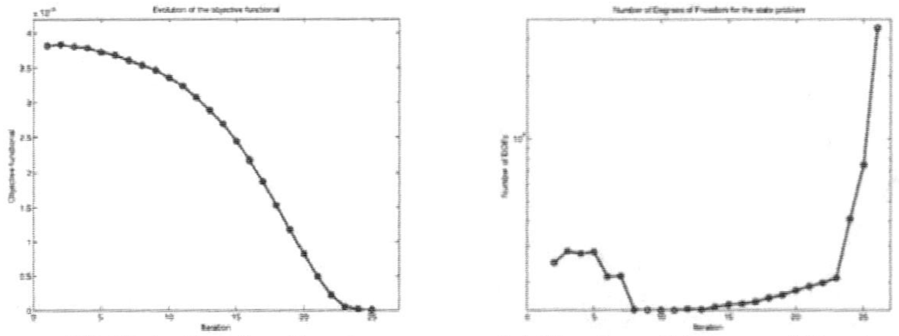

(d) Objective functional. (e) Number of Degrees of Freedom.

Figure 1: One inclusion in a circular domain. (a) Initial mesh. (b) Final mesh. (c) Initial configuration (dotted green), target inclusion (solid black) and reconstructed interface (dashed blue). (d) Objective functional. (e) Number of Degrees of Freedom.

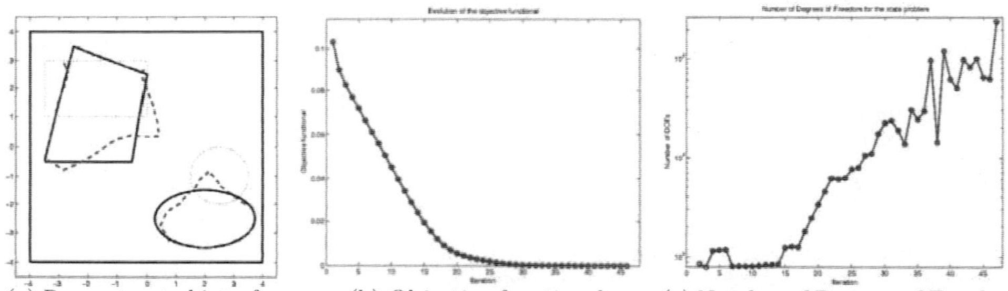

(a) Reconstructed interface. (b) Objective functional. (c) Number of Degrees of Freedom.

Figure 2: Two inclusions in a square domain. (a) Initial (dotted green), target (solid black) and reconstructed configuration (dashed blue). (b) Objective functional. (c) Number of DOFs.

dual flux problems to derive an upper bound of the error in the shape gradient is computationally expensive hence the ABVA may result in higher computing times than the classical BVA on fine meshes. Prompted by the above promising results, we are now investigating alternative procedures to estimate the error in order to rely only on the computation of local quantities.

References

[1] Oden J and Prudhomme S 2001 *Comput. Math. Appl.* **41** 735–756
[2] Banichuk N, Barthold F J, Falk A and Stein E 1995 *Struct. Optim.* **9** 46–51
[3] Morin P, Nochetto R, Pauletti M and Verani M 2012 *ESAIM: Control Optim. Calc. Var.* **18**(04) 1122–1149
[4] Kikuchi N, Chung K, Torigaki T and Taylor J 1986 *Comput. Method Appl. Mech. Eng.* **57** 67–89
[5] Schleupen A, Maute K and Ramm E 2000 *Struct. Multidisc. Optim.* **19** 282–302
[6] Kohn R V and Vogelius M 1987 *Comm. Pure Appl. Math.* **40** 745–777
[7] Céa J 1986 *Math. Model. Numer. Anal.* **20** 371–402
[8] Pantz O 2005 *C. R. Acad. Sci. Paris* **Ser. I** 333–337

Journal of Physics: Conference Series **657** (2015) 012005 doi:10.1088/1742-6596/657/1/012005

Two-time-scale inverse problems: formulation and solution

G Puel and D Aubry

Laboratoire MSSMat (CentraleSupélec – CNRS UMR8579 – Université Paris-Saclay), Grande Voie des Vignes, 92290 Châtenay-Malabry, France

E-mail: {guillaume.puel,denis.aubry}@centralesupelec.fr

Abstract. The aim here is to study two-time-scale models and their associated parameter identification. When it is possible to consider two well-separated time scales, and when the fast component of the applied loading is periodic, a periodic time homogenization scheme, similar to what exists in space homogenization, can be used to derive a homogenized model. A parameter identification process for this latter is then proposed, consisting in homogenizing with respect to time a classical identification strategy based on the use of adjoint state formulations; it is then applied to an academic example showing the benefits of such a strategy.

1. Introduction

As far as the numerical calculation of a time-dependent model is concerned, the question of the computational cost can be of utmost relevance, especially when the considered model deals with fast phenomena, which require the use of very small time steps, when compared with the length of the time interval of study. In order to drastically reduce the computational cost, a periodic time homogenization method, such as [1], can be used when well-separated time scales can be defined, and when the fast component of the applied loading is periodic. The resulting homogenized model is then cheaper computationally, for it can be solved using time steps related to the slow time scale only, whereas the fast time scale is taken into account in an average way in the homogenization scheme.

In order them to give accurate predictions, such time-homogenized models have to be compared with experimental data. The key point is to define an identification strategy able to deal with such models in such a way that the process remains cheap and efficient. The aim of this paper is to analyze on a specific academic example the different choices made throughout the whole identification process, and what is their impact on the identification results.

2. Periodic time homogenization method

Periodic time homogenization, as it was initially proposed in [1], can be seen as a transposition to time of the classical periodic space homogenization methods, such as the techniques described in [2, 3]. It consists in separating two time scales, a slow one t and a fast one $\tau = t/\xi$, by introducing, for every time-dependent quantity in the equations of the reference problem, an asymptotic expansion, in terms of the ratio ξ:

$$\alpha(t,\tau) = \alpha_0(t,\tau) + \xi\alpha_1(t,\tau) + \xi^2\alpha_2(t,\tau) + O(\xi^3) \tag{1}$$

Journal of Physics: Conference Series **657** (2015) 012005 doi:10.1088/1742-6596/657/1/012005

When this ratio ξ is very small, it is possible to consider that the two time scales are independent, and that any derivative with respect to time has to use the partial derivatives with respect to the two time scales:

$$\mathrm{d}_t\alpha = \partial_t\alpha + \frac{1}{\xi}\partial_\tau\alpha \tag{2}$$

where d_t, ∂_t, et ∂_τ stand for the total time derivative, the partial derivative with respect to the slow time and the partial derivative with respect to the fast time respectively. Moreover, if the applied loading has a component, which is periodic with respect to the fast time τ, it is reasonable to assume that every variable is quasiperiodic, meaning that it is periodic with respect to τ (with associated period T_f).

Using the asymptotic expansion (1) for every quantity in the reference equations, and balancing the different orders of ξ, the time-homogenized equations are determined by averaging over a fast period T_f the different quantities:

$$< \alpha > = \frac{\xi}{T_f}\int_0^{T_f/\xi} \alpha(t,\tau)\,\mathrm{d}\tau \tag{3}$$

allowing to separate slow-evolving phenomena from fast-time periodic components using the quasi-periodicity assumption. The residual quantities associated with this average are then denoted as $\alpha^* = \alpha - <\alpha>$, and depend on both time scales t and τ a priori: they usually have to verify a very simple problem (*e.g.* linear elastic). Eventually, these homogenized equations are solved relatively to the slow time scale only, by introducing at each slow time step the averaged influence of the fast cycles corresponding to the solution of the fast problem, allowing to solve for all the time-homogenized variables with a drastically reduced computational cost.

References about the periodic time homogenization still tend to be scarce, as showed in [4], even if the scope of applications can be quite large: see for example [5, 6, 7, 8, 9]. Recently, we focused on validating the method for different cases of simulations of structures withstanding fatigue loads with two periodic components:

- material fatigue with a viscoplastic law defining two hardening variables in [10];
- material fatigue with an isotropic damage law in [11];
- extension of the method to three different time scales in [4, 12];
- extension of the method to the case of resonant excitation in [4].

3. Gradient-based parameter identification with adjoint state formulations

In this section, a general identification strategy is proposed, which is relevant for both reference and time-homogenized models: it will thus be illustrated in this latter case in Section 4. The forward problem is considered as an implicit formulation with a vector function $\mathcal{F}$ considering $\mathbf{u}(t)$ over a time interval $[0, T]$:

$$\begin{aligned} &\mathcal{F}\left(\mathbf{u}(t), \mathrm{d}_t\mathbf{u}(t), \mathrm{d}_t^2\mathbf{u}(t), \mathbf{p}, t\right) = \mathbf{0} \\ &\mathbf{u}(0) = \mathbf{U}_0 \qquad \mathrm{d}_t\mathbf{u}(0) = \mathbf{V}_0 \end{aligned} \tag{4}$$

where d_t and d_t^2 are the first- and second-order time derivatives respectively. $\mathbf{U}_0$ and $\mathbf{V}_0$ stand for the initial conditions of the dynamic problem. Whereas $\mathbf{u}$ is the state vector of size N, composed of all the time-dependent degrees of freedom (DOFs) describing the studied problem, $\mathbf{p}$ stands for the vector containing the P scalar parameters associated with the differential equation (4).

The identification problem consists in finding the parameter vector $\mathbf{p}_{opt}$ such that the solution $\mathbf{u}(t; \mathbf{p}_{opt})$ of (4) obtained with the parameters $\mathbf{p}_{opt}$ is as close to the available experimental data as possible. These latter are indeed compared with the corresponding quantities $\mathbf{Au}(t; \mathbf{p})$, where $\mathbf{A}$ is a projection operator allowing to select, for each quantity, the closest DOF to the experimental measurement point. In order to use consistent notations, the corresponding experimental quantity is denoted $\mathbf{A}\mathbf{u}_{exp}(t)$; however, it does not mean that such a vector $\mathbf{u}_{exp}(t)$ actually exists.

The following misfit function is then introduced: it consists of a norm measuring the discrepancy between the quantities predicted with the forward model (4) and experimental data:

$$\mathcal{J}(\mathbf{p}) = \frac{1}{2}\int_0^T |\mathbf{A}(\mathbf{u}(t;\mathbf{p}) - \mathbf{u}_{exp}(t))|^2\,\mathrm{d}t + \frac{1}{2}|\mathbf{R}(\mathbf{p}-\mathbf{p}_0)|^2 \tag{5}$$

where $\mathbf{u}(t;\mathbf{p})$ satisfies Equation (4). The L^2-norm proposed here is completed with a Tikhonov regularization term, allowing to deal with the ill-posedness of the identification problem, by bounding the magnitude of the parameter vector $\mathbf{p}$ to be identified: this regularization term uses a vector $\mathbf{p}_0$ containing nominal values corresponding to a priori experience, and a diagonal weighing matrix $\mathbf{R}$. Eventually, the solution of the identification problem can be sought as the parameter vector minimizing the misfit function $\mathcal{J}(\mathbf{p})$:

$$\mathbf{p}_{opt} = \arg\min_{\mathbf{p}} \mathcal{J}(\mathbf{p}) \tag{6}$$

The determination of this minimum is achieved using gradient-based minimization methods, therefore the question of avoiding local minima by means of an appropriate regularization process should be carefully addressed. In some cases, rather than using the classical Tikhonov regularization term, the a priori experience can be introduced in some specific ways, as in [13]. Similarly, the fact of using a homogenized model in the parameter identification process can introduce a regularizing effect, just as explained in [14]. However, we will not address here this specific question, but rather focus on the identification process itself.

To estimate the gradient of the misfit function, we solve an adjoint state problem. A typical example in mechanical engineering is given in [15], where the parameters of an elastoplastic material law are identified with indentation tests. In the strategy proposed here, the generic form of the adjoint state problem is as follows:

$$\begin{aligned} &\nabla_{\mathbf{u}}\mathcal{F}^{\mathrm{T}}\mathbf{z} - \mathrm{d}_t\left(\nabla_{\mathrm{d}_t\mathbf{u}}\mathcal{F}^{\mathrm{T}}\mathbf{z}\right) + \mathrm{d}_t^2\left(\nabla_{\mathrm{d}_t^2\mathbf{u}}\mathcal{F}^{\mathrm{T}}\mathbf{z}\right) = \mathbf{A}^{\mathrm{T}}\mathbf{A}(\mathbf{u}-\mathbf{u}_{exp}) \\ &\left(\nabla_{\mathrm{d}_t^2\mathbf{u}}\mathcal{F}^{\mathrm{T}}\mathbf{z}\right)_{|t=T} = \mathbf{0} \qquad \left(\nabla_{\mathrm{d}_t\mathbf{u}}\mathcal{F}^{\mathrm{T}}\mathbf{z}\right)_{|t=T} = \mathbf{0} \end{aligned} \tag{7}$$

where $\nabla_{\mathbf{u}}\mathcal{F}$, $\nabla_{\mathrm{d}_t\mathbf{u}}\mathcal{F}$ and $\nabla_{\mathrm{d}_t^2\mathbf{u}}\mathcal{F}$ stand for the directional derivatives of $\mathcal{F}$ with respect to $\mathbf{u}$, $\mathrm{d}_t\mathbf{u}$ and $\mathrm{d}_t^2\mathbf{u}$ respectively. The adjoint state problem is then a time-backward differential equation with final conditions, and where the first-order sensitivities of the forward problem are concerned.

Once the adjoint state problem (7) is solved, it can be shown that the misfit function gradient with respect to the parameter vector $\mathbf{p}$ can be expressed as:

$$\nabla_{\mathbf{p}}\mathcal{J}(\mathbf{p}) = \mathbf{R}^{\mathrm{T}}\mathbf{R}(\mathbf{p}-\mathbf{p}_0) - \int_0^T \nabla_{\mathbf{p}}\mathcal{F}^{\mathrm{T}}\mathbf{z}(t)\,\mathrm{d}t \tag{8}$$

This specific way of estimating the misfit function gradient can be compared with a classical finite difference formula, such as the central finite difference scheme: in this latter case, when the parameter vector is of size P, the gradient calculation is obtained by evaluating the misfit function in $2P$ additional 'points', each couple of points corresponding to two symmetrical perturbations of the misfit function associated with each parameter in the vector $\mathbf{p}$. The resulting computational cost for each gradient evaluation consists of $2P$ solutions of the forward problem (4) and $2P$ time integral evaluations. By contrast, when the adjoint state solution is used, only two differential equation solutions are required: one for the forward problem (4) and one for the adjoint problem (7). The associated computational cost for each gradient evaluation is then 2 differential equation solutions, and P time integral evaluations. The resulting gain is as high as the number of parameters to be identified. Moreover, it is easier to control the accuracy of the gradient estimate with the adjoint state method than with finite difference formulas, for which the choice of the discretization steps has a strong influence on the final estimate.

4. Parameter identification of a model with two time scales

4.1. Forward reference and time-homogenized problems

Here an academic example is proposed to discuss the different steps of the parameter identification problem associated with a time-homogenized model. This latter consists of a straight bar of length L withstanding at one end a normal force f_s with two periodic components.

$$\partial_x \sigma + c_K \mathrm{d}_t \partial_x \sigma = \rho \mathrm{d}_t^2 u \tag{9}$$

$$u_{|x=0} = 0 \qquad (\sigma + c_K \mathrm{d}_t \sigma)_{|x=L} = f_\mathrm{s} \tag{10}$$

$$\sigma = E\left(\partial_x u - \varepsilon^\mathrm{p}\right) \tag{11}$$

$$\mathrm{d}_t \varepsilon^\mathrm{p} = \left(\frac{|\sigma|}{K}\right)^n \operatorname{sign} \sigma \tag{12}$$

where ∂_x is the partial space derivative, and zero initial conditions are assumed for the displacement u, $\mathrm{d}_t u$ and the plastic strain ε^p. K and n are the two parameters to be identified, using the measured displacement u_{exp} at $x = L$, whereas E, ρ and c_K are assumed known.

The time-homogenized equations come from the zeroth-order expression of the forward problem (9)-(10)-(11)-(12), where the fast-time average (3) has been previously applied:

$$\partial_x < \sigma_0 > = 0 \tag{13}$$

$$< u_0 >_{|x=0} = 0 \qquad < \sigma_0 >_{|x=L} = < f_\mathrm{s} > \tag{14}$$

$$< \sigma_0 > = E\left(\partial_x < u_0 > - \varepsilon_0^\mathrm{p}\right) \tag{15}$$

$$\partial_t \varepsilon_0^\mathrm{p} = \left\langle \left(\frac{|\sigma_0|}{K}\right)^n \operatorname{sign} \sigma_0 \right\rangle \tag{16}$$

where it has been assumed that $\rho L^2/(E T_f^2) \leq O(1)$ and $c_K/T_f \leq O(1)$, and where it can be shown that $\varepsilon_0^\mathrm{p}(x,t,\tau) = \varepsilon_0^\mathrm{p}(x,t)$. The fast-time average term associated with the evolution equation (16) is evaluated by means of a numerical integration formula, such as the trapezoidal rule, and uses the solution of the residual system defined as the difference between the reference problem and the zeroth-order time-homogenized system (13)-(14)-(15)-(16).

4.2. Identification strategy

The first step consists in describing, through the misfit function, the discrepancy between the time-homogenized model's predictions and experimental data: indeed, on the one hand, the model has been solved on slow time steps only, whereas, on the other hand, the experimental data can be available on a much finer scale.

The most efficient choice in terms of computation cost is to use time-homogenized quantities in the misfit function, since it allows to address the time integral on slow time steps only:

$$\mathcal{J}_0(K,n) = \frac{1}{2}\int_0^T |< u_0 > (L,t;K,n) - < u_{exp} > (t)|^2 \,\mathrm{d}t \tag{17}$$

Whereas $< u_0 > (L,t;K,n)$ is the solution of the zeroth-order time-homogenized forward problem (13)-(14)-(15)-(16), $< u_{exp} > (t)$ stands for the corresponding experimental quantity, which is obtained by fast-averaging the experimental data for each slow time step t_k of the time-homogenized displacement:

$$< u_{exp} > (t_k) = \frac{1}{T_f} \int_{t_k}^{t_k+T_f} u_{exp}(t) \, \mathrm{d}t \tag{18}$$

The gradient of this misfit function is evaluated as in Section 3, using the solution of an adjoint state problem. This latter is as follows, with $\lambda_0(x,t)$ only:

$$\begin{aligned} \partial_t \lambda_0 &= < u_0 >_{|x=L} - < u_{exp} > \\ \lambda_{0\,|t=T} &= 0 \end{aligned} \tag{19}$$

and it can be shown that it corresponds to the zeroth-order time-homogenized version of the time-backward adjoint state problem associated with the identification process for the reference problem. Indeed, Figures 1 and 2 show the comparison between the two corresponding adjoint state solutions for the identification process detailed previously: the adjoint state solution corresponding to the time-homogenized problem is homogeneous along the bar, and is very close to the (homogeneous) fast-time average of the (heterogeneous) adjoint state solution associated with the reference problem. This equation can be solved using the slow time steps t_k only, which allows to derive the solution in a way as efficient as for the time-homogenized forward solution (13)-(14)-(15)-(16).

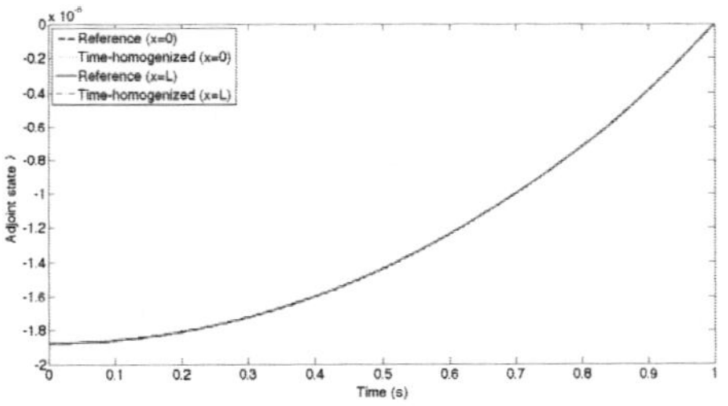

Figure 1. Adjoint state solutions for reference (in red) and time-homogenized (in blue) problems, at $x = 0$ and $x = L$.

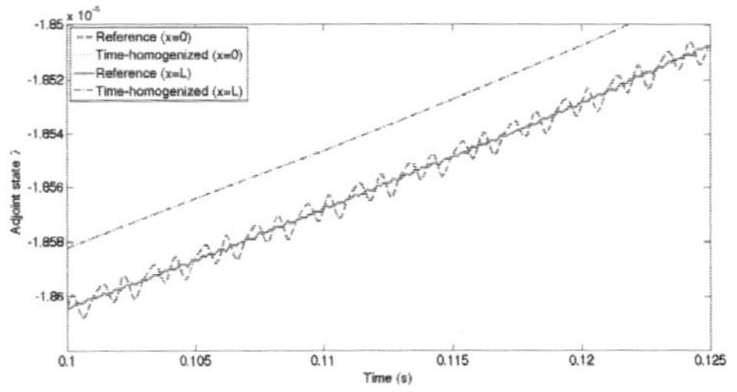

Figure 2. Zoom of Figure 1 (same color conventions).

The misfit function's gradient then consists of the two following partial derivatives:

$$\frac{\partial \mathcal{J}_0}{\partial K} = \int_0^T \int_0^L \left\langle \frac{n}{K} \left(\frac{E \left| \partial_x u_0 - \varepsilon_0^{\mathrm{p}} \right|}{K} \right)^n \mathrm{sign} \left(\partial_x u_0 - \varepsilon_0^{\mathrm{p}} \right) \right\rangle \lambda_0 \, \mathrm{d}x \, \mathrm{d}t \tag{20}$$

$$\frac{\partial \mathcal{J}_0}{\partial n} = - \int_0^T \int_0^L \left\langle \left(\frac{E \left| \partial_x u_0 - \varepsilon_0^{\mathrm{p}} \right|}{K} \right)^n \mathrm{sign} \left(\partial_x u_0 - \varepsilon_0^{\mathrm{p}} \right) \log \left(\frac{E \left| \partial_x u_0 - \varepsilon_0^{\mathrm{p}} \right|}{K} \right) \right\rangle \lambda_0 \, \mathrm{d}x \, \mathrm{d}t \tag{21}$$

Once again, it can be shown that these relations correspond to the zeroth-order time-homogenized estimates of the two misfit function's gradient components obtained for the identification problem associated with the reference problem. Actually, this is a result that already occurs in periodic space homogenization, as shown for example in [16].

In order to evaluate how the identification process performs, synthetic data $u_{exp}(t)$ are created by solving the reference forward problem (13)-(14)-(15)-(16), using $K_{exp} = 100 \cdot 10^6$ SI units and $n_{exp} = 10$ as parameter values. The bar withstands a two-periodic loading with $F = 0.05$ Hz and $F/\xi = 500$ Hz. $K_0 = 50 \cdot 10^6$ SI units and $n_0 = 5$ are chosen as initial parameter values for the identification process, which is based on an interior-reflective Newton method [17] in order to minimize the misfit function $\mathcal{J}^0$. Since we do not want to address specifically the question of regularization, no Tikhonov term is added. Figure 3 shows the comparison between the identified model ($K_{id} = 100.89 \cdot 10^6$ SI units and $n_{id} = 9.94$) and the synthetic reference, more precisely the variations of the longitudinal plastic strain, which is not directly observable. When compared with what is obtained when the inverse problem related to the reference problem (9)-(10)-(11)-(12) is considered, the computational cost associated with the identification process is significantly reduced (10^5 time steps instead of 10^7 here).

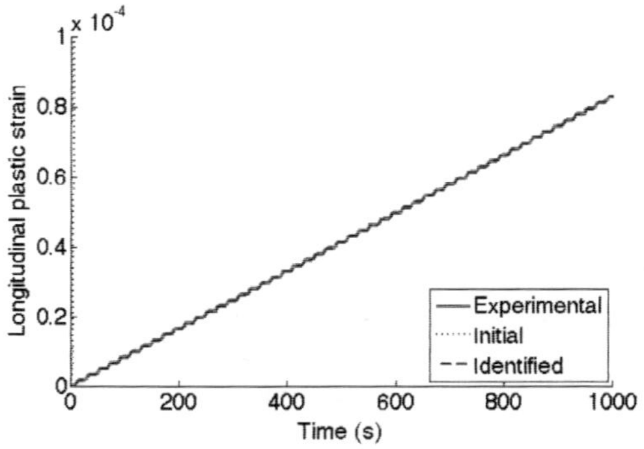

Figure 3. Parameter identification of a time-homogenized model: experimental (in red) and predicted (in blue) longitudinal plastic strain.

5. Conclusion

Here we have proposed a first preliminary study of a two-time-scale parameter identification process, using time-homogenized models: the adaptation of a classical identification strategy based on an adjoint state formulation to estimate the misfit function's gradient can be used in this specific framework: this leads, on the proposed example, to the determination of the time-homogenized counterpart of the adjoint solution associated with the reference identification problem. Despite its simplicity, the academical example studied here showed the relevance of the strategy and its reduced computational cost: these results can be viewed as a first step before dealing with more complex cases of study.

References

[1] Guennouni T and Aubry D 1986 *Comptes rendus de l'Académie des sciences. Série II* **303** 1765–1768
[2] Bensoussan A, Lions J L and Papanicolaou G 1978 *Asymptotic Analysis for Periodic Structures* (Burlington, MA: Elsevier)
[3] Sanchez-Palencia E 1980 *Non-homogeneous media and vibration theory* (Berlin Heidelberg New York: Springer-Verlag)
[4] Puel G and Aubry D 2014 *International Journal for Multiscale Computational Engineering* **12** 291–318
[5] Yu Q and Fish J 2002 *Computational Mechanics* **29** 199–211
[6] Shishkina E V, Blekhman I I, Cartmell M P and Gavrilov S N 2008 *Nonlinear Dynamics* **54** 313–331
[7] Lévy A, Le Corre S, Poitou A and Soccard E 2011 *International Journal for Multiscale Computational Engineering* **9** 53–72
[8] Dontsov E V and Guzina B B 2012 *Journal of the Mechanics and Physics of Solids* **60** 1791–1813
[9] Papon A, Yin Z Y, Riou Y and Hicher P Y 2013 *International Journal for Numerical and Analytical Methods in Geomechanics* **37** 1470–1491
[10] Aubry D and Puel G 2010 *Procedia Engineering* **2** 787–796
[11] Devulder A, Aubry D and Puel G 2010 *Computational Mechanics* **45** 637–646
[12] Puel G and Aubry D 2012 *European Journal of Computational Mechanics* **21** 312–324
[13] Puel G, Bourgeteau B and Aubry D 2013 *Mechanical Systems and Signal Processing* **36** 354–369
[14] Chung C B and Kravaris C 1988 *Inverse Problems* **4** 973–994
[15] Constantinescu A and Tardieu N 2001 *Inverse Problems in Engineering* **9** 19–44
[16] Kesavan S and Saint Jean Paulin J 1997 *SIAM Journal on Control and Optimization* **35** 1557–1573
[17] Coleman T F and Li Y 1994 *Mathematical Programming* **67** 189–224

5th International Workshop on New Computational Methods for Inverse Problems IOP Publishing
Journal of Physics: Conference Series **657** (2015) 012006 doi:10.1088/1742-6596/657/1/012006

Numerical solution of a nonlinear least squares problem in digital breast tomosynthesis

G. Landi[1], E. Loli Piccolomini[1], J. G. Nagy[2]
[1]Department of Mathematics, University of Bologna
[2]Department of Mathematics and Computer Science, Emory University
E-mail: germana.landi@unibo.it

Abstract. In digital tomosynthesis imaging, multiple projections of an object are obtained along a small range of different incident angles in order to reconstruct a pseudo-3D representation (i.e., a set of 2D slices) of the object. In this paper we describe some mathematical models for polyenergetic digital breast tomosynthesis image reconstruction that explicitly takes into account various materials composing the object and the polyenergetic nature of the x-ray beam. A polyenergetic model helps to reduce beam hardening artifacts, but the disadvantage is that it requires solving a large-scale nonlinear ill-posed inverse problem. We formulate the image reconstruction process (i.e., the method to solve the ill-posed inverse problem) in a nonlinear least squares framework, and use a Levenberg-Marquardt scheme to solve it. Some implementation details are discussed, and numerical experiments are provided to illustrate the performance of the methods.

1. Introduction

Digital tomosynthesis is an emerging technique in digital radiography with the aim of observing a 3D object with low dose radiation [1, 2]. It acquires 2D image projections of a 3D object for varying incident angles. From the set of the 2D acquired images, the reconstruction algorithms should be able to reconstruct any number of slices of the 3D object. In order to make the exam less invasive, the angular range is limited, which requires special consideration in the reconstruction algorithms. For this reason, iterative algorithms are preferable to standard Filtered Backprojection.

In this paper, we consider a reconstruction model taking into account the polyenergetic nature of the x-ray beam and the different materials composing the object. In [3] the authors present a breast tomosynthesis polyenergetic model but they consider only the percentage of one material, while, in [4], the polyenergetic model is extended to the multi-material case by considering the linear attenuation coefficients of each material. The results reported in [4] show that the multimaterial polyenergetic model produces images of improved quality with respect to those obtained by the classical FBP method, reducing noise and beam hardening artifacts. In the following, we present a multi-material polyenergetic model based on the mass attenuation coefficients of the materials composing the object and we reconstruct pseudo-3D images for the mass coefficients of each material. The linear attenuation coefficient of a material represents its attenuation property but it linearly depends on the density (for example, the vapor water and the liquid water have different linear attenuation coefficient values), while the mass attenuation coefficient of a material represents its attenuation property per unit mass.

For this reason, the mass attenuation coefficients have great interest in clinical radiography. In particular, in breast tomosynthesis, the object is composed of air, glandular tissue, adipose tissue and microcalcifications.

The noise on the projection data is a mixture of Poisson noise, coming from x-ray scattering, and Gaussian noise introduced in the digital registration. Since it is reported in [5] that the quality of the reconstructed image is not severely influenced by the noise model, in this paper we suppose to have only Gaussian noise.

The resulting nonlinear least squares problem is large size and ill-posed. Hence, regularization is necessary to compute a good solution. We apply the Levenberg-Marquardt (LM) method that, in our experiments, has shown good performance in terms of reconstructed image quality and computational time. The numerical results presented here for a simulated breast tomosynthesis imaging problem show that the approach is promising and it should be further investigated with more numerical tests on real data with different numerical methods.

In sections 2 and 3 we describe the considered multi-material polyenergetic model and the LM method, respectively. In section 4 we provide some numerical results to illustrate the performance of the numerical methods, and in 5, we present some conclusions.

2. The reconstruction model

Traditionally, the reconstruction models assume, for simplicity, that the source ray is monochromatic, even if the x-ray photons are emitted at different energies in a defined range. Ignoring the energy dependence in the model produces the so called "beam hardening" artifacts in the reconstructed image. In [3] the authors proposed a model accounting for a polyenergetic x-ray. If we consider a 3D object discretized in N_v voxels, a system detector with N_p pixels moving through N_θ angles, the projection in the pixel i for the incident angle θ is expressed as:

$$b_i^\theta = \sum_{e=1}^{N_e} s_e \exp\left(-\sum_{j=1}^{N_v} a_{i,j}^\theta \mu_{j,e} \right) + \eta_i^\theta \quad i = 1, \dots N_p, \quad \theta = 1, \dots N_\theta, \tag{1}$$

where s_e is the term accounting for the energy fluency, i.e. the product of x-ray energy with the number of incident photons at that energy with N_e discrete energy levels, $a_{i,j}^\theta$ is the length of the incident ray passing through the voxel j and contributing to pixel i, $\mu_{j,e}$ is the linear attenuation coefficient in voxel j at the energy e and η_i^θ represents measurement errors.

In [4] the model (1) was modified by supposing the object constituted of N_m different materials, each with a specific linear attenuation coefficient $c_{m,e}$, so that the the linear attenuation coefficients $\mu_{j,e}$ can be approximated as:

$$\mu_{j,e} = \sum_{m=1}^{N_m} w_{j,m} c_{m,e} \quad \text{with} \quad \sum_{m=1}^{N_m} w_{j,m} = 1. \tag{2}$$

The unknowns $w_{j,m}$ are the weights of the linear attenuation coefficients $c_{m,e}$ of each material and they depend on the percentage of the mth material in voxel j at the x-ray energy e. Hence equation (1) can be written in terms of the $w_{j,m}$ as:

$$b_i^\theta = \sum_{e=1}^{N_e} s_e \exp\left(-\sum_{j=1}^{N_v} a_{i,j}^\theta \sum_{m=1}^{N_m} w_{j,m} c_{m,e} \right) + \eta_i^\theta \quad i = 1, \dots N_p, \quad \theta = 1, \dots N_\theta. \tag{3}$$

In the following, we consider the mass attenuation coefficients $\delta_{j,e}$ for the voxel j at the x-ray energy e that are related to the linear attenuation coefficients $\mu_{j,e}$ by:

$$\mu_{j,e} = \rho_j \delta_{j,e} \tag{4}$$

where ρ_j is the density of the composite materials in voxel j. Similarly to equation (2), we can write $\delta_{j,e}$ as a linear combination of the mass attenuation coefficients $d_{m,e}$ of the mth material at the x-ray energy e through some weights $\widetilde{w}_{j,m}$:

$$\delta_{j,e} = \sum_{m=1}^{N_m} \widetilde{w}_{j,m} d_{m,e} \quad \text{with} \quad \sum_{m=1}^{N_m} \widetilde{w}_{j,m} = 1. \tag{5}$$

Moreover, the densities ρ_j can be expressed as:

$$\rho_j = \frac{1}{\displaystyle\sum_{m=1}^{N_m} \frac{\widetilde{w}_{j,m}}{r_m}} \tag{6}$$

where r_m is the density of the mth material. By substituting (4), (5) and (6) in (1) we obtain the expression of the projection in pixel i at the angle θ as a function of the mass attenuation coefficients:

$$b_i^\theta = \sum_{e=1}^{N_e} s_e \exp\left(-\sum_{j=1}^{N_v} a_{i,j}^\theta \frac{\displaystyle\sum_{m=1}^{N_m} \widetilde{w}_{j,m} d_{m,e}}{\displaystyle\sum_{m=1}^{N_m} \frac{\widetilde{w}_{j,m}}{r_m}} \right) + \eta_i^\theta, \quad i = 1, \dots N_p, \quad \theta = 1, \dots N_\theta. \tag{7}$$

If we introduce the $N_v \times N_m$ matrix $\widetilde{\mathbf{W}}$ with entries $\widetilde{w}_{j,m}$, the $N_p N_\theta \times N_v$ matrix $\mathbf{A}$ with entries $a_{i,j}^\theta$, the vectors $\mathbf{b}$ and $\boldsymbol{\eta}$ of length $N_p N_\theta$ obtained by storing the projections b_i^θ and the noise elements η_i^θ, respectively, and the vector $\mathbf{s}$ of length N_e containing the values s_e then, we can write equation (7) in matrix-vector form as:

$$\widetilde{\mathbf{r}}(\widetilde{\mathbf{W}}) = 0 \tag{8}$$

where $\widetilde{\mathbf{r}}$ is a function from $\mathbb{R}^{N_v N_m}$ to $\mathbb{R}^{N_p N_\theta}$ defined as

$$\widetilde{\mathbf{r}}(\widetilde{\mathbf{W}}) := \mathbf{b} - \left(\exp\left(-\mathbf{A} \left(\frac{\widetilde{\mathbf{W}}\mathbf{D}}{\widetilde{\mathbf{W}}\mathbf{R}} \right) \right) \mathbf{s} + \boldsymbol{\eta} \right), \quad \mathbf{D}_{m,e} := d_{m,e}, \quad \mathbf{R}_{m,e} := \frac{1}{r_m} \tag{9}$$

where $\mathbf{D}$ and $\mathbf{R}$ have size $N_m \times N_e$. The $\exp(\cdot)$ function and the division operation are intended to be performed element-wise, while the arguments of the exponential function is computed with matrix-matrix and matrix-vector operations. In practice, the x-ray energies can be estimated by using the well-known spectra models [6], the matrix $\mathbf{D}$ and $\mathbf{R}$ can be obtained by taking x-ray transmission measurements of objects that have known dimension, density and material composition, while the noise mean and variance can be estimated through a calibration process. The unknown weights $\widetilde{w}_{j,m}$ of the mass attenuation coefficients can be obtained by solving the inverse problem (8).

3. Numerical solution of the inverse problem

For the numerical solution of the inverse problem (8), we employ the widely used least squares approach requiring to solve the constrained problem:

$$\begin{aligned} &\min_{\widetilde{\mathbf{W}}} \frac{1}{2} \|\widetilde{\mathbf{r}}(\widetilde{\mathbf{W}})\|^2 \\ &\text{s.t.} \quad \mathbf{W}\mathbf{1}_{N_m} = \mathbf{1}_{N_v} \end{aligned} \tag{10}$$

where $\mathbf{1}_N$ is the vector of length N whose elements are all 1. This isn't the only possible approach for solving the inverse problem (8). For example, a Bayesian approach could be considered to model the uncertainty in the estimated matrices $\mathbf{D}$ and $\mathbf{R}$ or to strengthen the Poissonian component of the noise.

Upon variable substitution for $\widetilde{w}_{j,1} = 1 - \sum_{m=2}^{N_m} \widetilde{w}_{j,m}$, problem (10) becomes the unconstrained nonlinear least squares problem

$$\min_{\mathbf{W}} \frac{1}{2}\|\mathbf{r}(\mathbf{W})\|^2 \tag{11}$$

where, using Matlab-like notation, $\mathbf{W} = \widetilde{\mathbf{W}}(:, 2 : N_m)$ and

$$\mathbf{r}(\mathbf{W}) := \mathbf{b} - \left(\exp\left(-\mathbf{A} \begin{pmatrix} [1 - \sum_{m=1}^{N_m-1} \mathbf{W}(:, m), \quad \mathbf{W}]\mathbf{D} \\ [1 - \sum_{m=1}^{N_m-1} \mathbf{W}(:, m), \quad \mathbf{W}]\mathbf{R} \end{pmatrix} \right) \mathbf{s} + \boldsymbol{\eta} \right). \tag{12}$$

Usually, problem (11) is ill-posed and the Jacobian matrix $\mathbf{J}$ of the residual vector $\mathbf{r}(\mathbf{W})$ is severely ill-conditioned. In order to account for the ill-conditioning of $\mathbf{J}$, for the numerical solution of (11), we consider a globally convergent modified LM method [7] defined by the general iteration:

$$\mathbf{W}_{k+1} = \mathbf{W}_k + \alpha_k \mathbf{p}_k \tag{13}$$

where $\mathbf{p}_k$ satisfies

$$(\mathbf{J}_k^T \mathbf{J}_k + \lambda_k \mathbf{I})\mathbf{p}_k = -\mathbf{J}_k^T \mathbf{r}_k \tag{14}$$

and the step-length α_k is determined by an Armijo-like procedure with initial unit step-length [7]. The scalar λ_k is chosen according to the following criterion:

$$\lambda_k = \max(c_1, \min(c_2, \|\mathbf{J}_k^T \mathbf{r}_k\|)), \quad \lambda_0 = c_2 \tag{15}$$

with $0 < c_1 < \lambda_k < c_2$. This criterion prevents severe ill-conditioning in the matrix $\mathbf{J}_k^T \mathbf{J}_k + \lambda_k \mathbf{I}$ and accelerates convergence when the residuals are small. Observe that the proposed criterion follows the original Levenberg-Marquardt strategy of selecting λ_k directly instead of using a trust-region approach to manipulate the choice of λ_k. Global convergence is ensured by the line-search procedure.

4. Numerical results

In this section, we present some preliminary results obtained for a simulated breast imaging problem. The experiments were performed using Matlab 7.14 (R2012a) on a Sun Fire V40z server consisting of four 2.4GHz AMD Opteron 850 processors with 16GB RAM. In our test problem, we considered one simulated 3D phantom object of size $N_v = 64 \times 64 \times 7$, made of four ellipses consisting of a tissue mixture with varying percentages of glandular and adipose tissue and with homogeneous background made of a mixture of 50% adipose and 50% glandular tissue. The central slice of the exact 3D weights image of the mass attenuation coefficients of the glandular material is shown in figure 1. Observe that the weights of the mass attenuation coefficients of the adipose material can be obtained from those of the glandular material by using the equation $\sum_{m=1}^{N_m} \widetilde{w}_{j,m} = 1$. The matrix $\mathbf{A}$ was obtained by considering $N_\theta = 15$ equispaced projection angles from $-17°$ to $17°$ and, for each projection angle, the ray trace matrix was computed using a cone beam model with Siddon's algorithm for ray tracing [8]. Finally, $N_e = 37$ different levels of energy were considered from 10 keV to 28 keV, in 0.5 keV steps. Gaussian noise was added to the 15 projection images to simulate experimental errors. We define the noise level, nl, as the ratio between $\|\boldsymbol{\eta}\|_2$ and $\|\boldsymbol{b} - \boldsymbol{\eta}\|_2$; the noise levels 10^{-3} and 10^{-4} were used in our experiments. We note that lower is the noise level, lower is the noise on the image; sometimes the Signal to Noise ratio (SNR) is used to measure the noise intensity that, conversely, is higher when the noise is greater. The nonlinear least squares problem (11) was solved by the LM method (13)-(14) where central finite differences were used to approximate the Jacobian matrix $\mathbf{J}_k$ and the values of the parameters c_1 and c_2 were respectively fixed at 10^{-2} and 10^3 for $nl = 10^{-3}$, and 10^{-2} and 1 for $nl = 10^{-4}$. The linear system (14) was solved by using the Cholesky decomposition. The LM method exhibits a semi-convergent behavior [9] and the iteration plays the role of the regularizing parameter. Since, in this work, we are mainly interested in analyzing the capability of the reconstruction approach (11) to compute the weights of the mass attenuation coefficients, in figure 1, we present the central slice of the restored 3D weights images of the glandular mass attenuation coefficients for the glandular material with the lowest relative error. Figure 2 depicts the horizontal line slice through the center of the weights images of figure 1. For $nl = 10^{-3}$, the optimal error value 0.0408 was obtained at the 9th iteration; for $nl = 10^{-4}$, the optimal error value 0.0139 was achieved at the 9th iteration. Figure 3 shows the restored weights images in the YZ plane and figure 4 depicts the corresponding line slice through their center. Finally, figure 5 presents the iteration history of the relative error of the unknown weights $\mathbf{W}_k$ (left) and the objective function (right) for $nl = 10^{-3}$. The plot on the left illustrates the semi-convergence property of the LM method, while the plot on the right shows that the objective function becomes nearly flat in the neighborhood of the iteration where the lowest relative error value is obtained. This behavior has been observed in all the performed numerical tests.

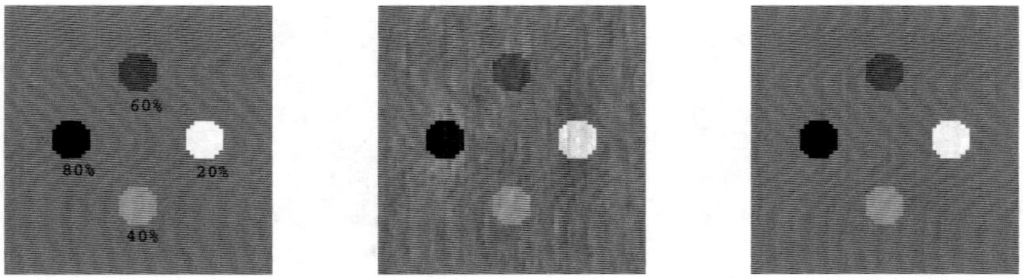

Figure 1: Central slice of the restored 3D weights images of the mass attenuation coefficients for the glandular material. From left to right: exact slice, reconstruction slice for $nl = 10^{-3}$ and for $nl = 10^{-4}$.

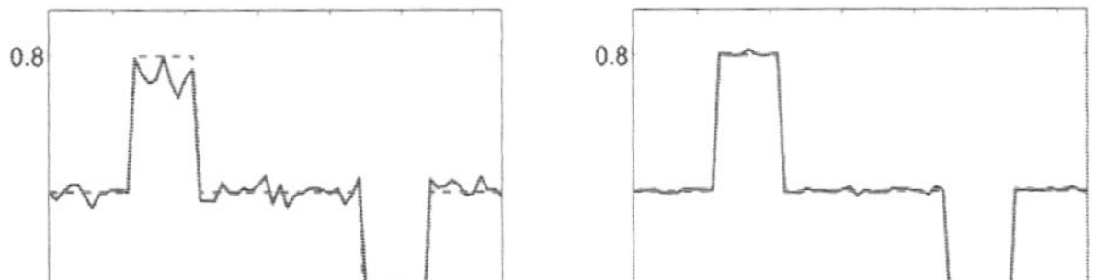

Figure 2: Horizontal profile through the center of the central slice of the 3D weights images of the mass attenuation coefficients for the glandular material for $nl = 10^{-3}$ (left) and $nl = 10^{-4}$ (right).

5. Conclusions

In this work we have proposed an approach to digital breast tomosynthesis where information about both the different materials composing the object and the different energy levels of the

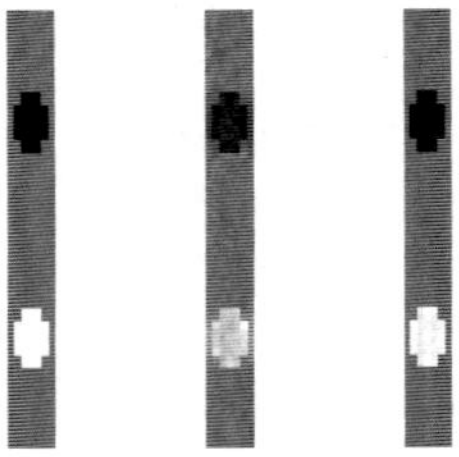

Figure 3: Central slice of the glandular weights images in the YZ plane. From left to right: exact slice, reconstruction slice for $nl = 10^{-3}$ and $nl = 10^{-4}$.

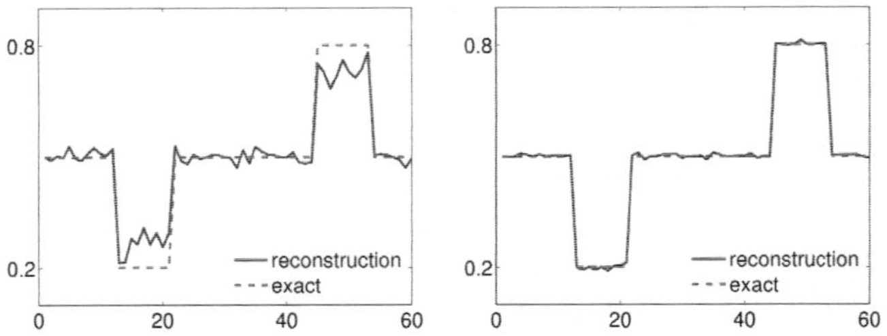

Figure 4: Profile through the center of the central slice of the 3D weights images of the mass attenuation coefficients for the glandular material in the YZ plane for $nl = 10^{-3}$ (left) and $nl = 10^{-4}$ (right).

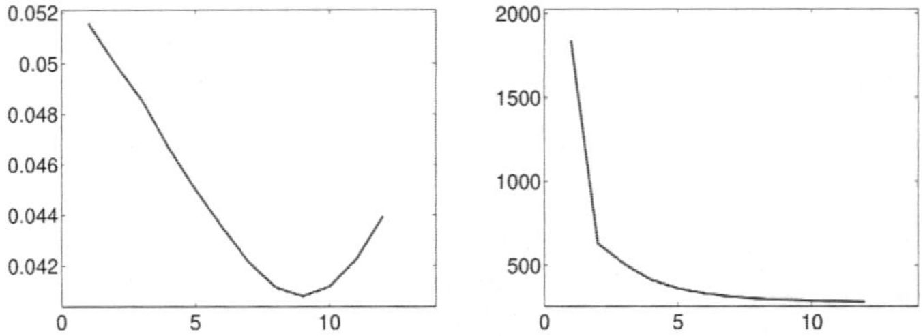

Figure 5: Objective function (left) and relative error (right) histories for $nl = 10^{-3}$.

x-ray beam is incorporated in the reconstruction model. In particular, we have examined the inverse problem of determining the mass attenuation coefficients of the various materials at the different x-ray energy levels and we have proposed a Levenberg-Marquardt-type method for its solution. Regularization has been enforced by early termination of the iteration. Initial results of simulated test problems show the potential of the proposed approach, since the method well separate the two considered materials and it seems robust with respect to the noise, although there is still research to be done regarding the regularized inversion method. Moreover, we think that the first step should be to develop a parallel implementation of the proposed approach in order to be able to handle large-scale clinical data with thousands of unknowns.

References

[1] Grant D 1972 *IEEE Trans. Biom. Eng.* **19** 20–28
[2] Gobbins J and Godfrey D 2003 *Phys. Med. Biol.* **48** R65–R106
[3] Chung J, Nagy J and Sechopoulus I 2010 *SIAM J. Imag. Sci.* **3** 133
[4] Bustamante V, Nagy J, Feng S and Sechopoulus I 2013 *SIAM J. Scient. Comput.* **135** S192–S208
[5] Lu Y, Chan H, Wei J, Goodsitt M, Carson P, Hadjiiski L, Schmitz A, Eberhard J and Claus B 2006 *Med. Phys.* **33** 3781–3795
[6] Cranley K, Gilmore B, Fogarty G and Desponds L 1997 Catalogue of diagnostic x-ray data and other data Tech. Rep. 78. Institute of Physics and Engineering in Medicine (IPEM) York
[7] Nocedal J and Wright S J 2006 *Numerical Optimization* (New York: Springer)
[8] Siddon R L 1985 *Med. Phys.* **12** 252–255
[9] Kaltenbacher B, Neubauer A and Scherzer O, 2008 *Iterative Regularization Methods for Nonlinear Ill-Posed Problems* (Walter de Gruyter, Berlin)

Journal of Physics: Conference Series **657** (2015) 012007 doi:10.1088/1742-6596/657/1/012007

Coupling Modified Constitutive Relation Error, Model Reduction and Kalman Filtering Algorithms for Real-Time Parameters Identification

Basile Marchand[1], Ludovic Chamoin[1], Christian Rey[2]

[1] LMT (ENS Cachan/CNRS/Paris-Saclay University), 61 Avenue du Président Wilson, 94230 Cachan, France

[2] SAFRAN TECH, 1 Rue Geneviève Aubé, CS 80112, 78772 Magny les Hameaux, France

E-mail: {marchand,chamoin}@lmt.ens-cachan.fr, christian.rey@safran.fr

Abstract. In this work we propose a new identification strategy based on the coupling between a probabilistic data assimilation method and a deterministic inverse problem approach using the modified Constitutive Relation Error energy functional. The idea is thus to offer efficient identification despite of highly corrupted data for time-dependent systems. In order to perform real-time identification, the modified Constitutive Relation Error is here associated to a model reduction method based on Proper Generalized Decomposition. The proposed strategy is applied to two thermal problems with identification of time-dependent boundary conditions, or material parameters.

1. Introduction

The Dynamic Data-Driven Application Systems (DDDAS) concept has received an increasing interest during the last decade, in particular in the Computational Mechanics community [6]. The main idea is to create a feedback loop between a real system and its numerical model, in order to: (i) control the evolution of the system using model predictions; (ii) update model parameters using data measured on the physical system. A potential application, which is the final target of this work, is structural health monitoring with real-time identification and control of damage evolution in composite materials [13].

In this work, we focus on the identification step and model updating procedure; this requires solving an inverse problem which is usually ill-posed. In order to investigate inverse problems many approaches are possible : (i) deterministic approaches based on the definition of a cost function associated to a regularization method; (ii) stochastic approaches using Bayesian inference. The approach we propose here is based on a coupling between stochastic and deterministic approaches. More specifically, we introduce the modified Constitutive Relation Error (mCRE) which is an energy based method [11] and the Kalman filtering which is a bayesian data assimilation method [10]. On the one hand, Kalman filtering enables effective data assimilation and prediction of a dynamic system evolution from incomplete information; it can be extended to non-linear systems. On the other hand, mCRE is a robust and powerful tool for complex model identification; leaning on energy functionals as well as duality and convexity properties, it has the ability to identify model parameters from highly corrupted data [2]. In order to reach the real-time feature of the inverse method, mCRE is here associated with

Journal of Physics: Conference Series **657** (2015) 012007 doi:10.1088/1742-6596/657/1/012007

reduced order modelling based on Proper Generalized Decomposition [5, 4]. The proposed data assimilation strategy is applied to two thermal problems with real-time identification/updating of evolving boundary conditions, or material parameters.

2. Problem setting

We consider a thermal evolution of a body $\Omega \subset \mathbb{R}^d$ over the time interval $I_t = [0, T]$. The boundary of Ω, denoted $\partial\Omega$, is split in $\partial\Omega_u$ (where Dirichlet boundary conditions $u = u^d$ are applied) and $\partial\Omega_{\mathbf{q}}$ (where Neumann boundary conditions $\mathbf{q} \cdot \mathbf{n} = q^d$ are applied). We suppose that $\partial\Omega_u$ and $\partial\Omega_{\mathbf{q}}$ are non-overlapping boundaries. In a direct approach, the thermal problem we consider can be written under the following weak form :

$$\mathcal{B}(u, v) = \mathcal{L}(v) \;\; \forall v \in \mathcal{V} \otimes \mathcal{T}$$

with admissibility spaces $\mathcal{V} = \left\{u \in H^1(\Omega) \setminus u = u^d \, \forall \mathbf{x} \in \partial\Omega_u\right\}$ and $\mathcal{T} = \left\{u \in L^2(I_t) \setminus u = u^0 \text{ for } t = 0\right\}$. The operators are defined by $\mathcal{B}(u, v) = \int_{I_t} \int_\Omega (c\dot{u}v + \mathcal{K}\nabla u \nabla v) d\mathbf{x}dt$ and $\mathcal{L}(v) = \int_{I_t} \int_\Omega f v d\mathbf{x}dt + \int_{I_t} \int_{\partial\Omega_{\mathbf{q}}} q^d v d\mathbf{x}dt$ with u the temperature field, c the thermal capacity, $\mathcal{K}$ the thermal conductivity and f a source term.

In the context of inverse problems, we search to identify a set $\mathbf{p} \in \mathbb{R}^{n_p}$ of model parameters related for instance to material behavior. The problem we search to solve is then to find $\mathbf{p}$ such that :

$$\mathcal{G}(s, v; \mathbf{p}) = 0 \;\; \forall v \in \mathcal{V} \otimes \mathcal{T} \tag{1}$$

where s denotes observations data realized on the physical system and operator $\mathcal{G}$ is defined from the direct problem and introducing a measurement error term. Classically the problem (1) corresponds to a constrained minimization problem. However, according to DDDAS paradigm where observations are not known for the whole time interval but sequentially assimilated, we reformulate the problem under the form of a dynamical system. The dynamical system is given under space and time discrete form in (2) where we use a mesh of n nodes for space discretization and a time discretization composed to n_t steps; moreover we consider m measurement nodes.

$$\begin{cases} \mathbf{u}^{(k+1)} = \mathcal{M}^{(k)}\mathbf{u}^{(k)} + \mathbf{e}_M^{(k)} \\ \mathbf{s}^{(k)} = \mathcal{H}^{(k)}\mathbf{u}^{(k)} + \mathbf{e}_S^{(k)} \end{cases} \tag{2}$$

$\mathbf{u}^{(k)} \in \mathbb{R}^n$ is the nodal temperature field at time $t^{(k)}$, $\mathbf{e}_M^{(k)} \in \mathbb{R}^n$ is a modeling error term, $\mathbf{s}^{(k)} \in \mathbb{R}^m$ is the nodal values of observed data, and $\mathbf{e}_S^{(k)} \in \mathbb{R}^m$ is an observation error reflecting sensor sensitivity. Operators $\mathcal{M}^{(k)} : \mathbb{R}^n \to \mathbb{R}^n$ and $\mathcal{H}^{(k)} : \mathbb{R}^n \to \mathbb{R}^m$ are respectively model (or transition) and observation operators.

3. Proposed model updating strategy

3.1. Classical way based on Kalman filtering

The Kalman filtering is a well-known data assimilation method [10], applicable for the resolution of inverse problems and widely used in mechanical identification problems (see for example [3]). It can be seen as a Bayesian filter in the Gaussian particular case. The Bayesian filter principle is to introduce the two following hypotheses in the Bayes theorem : (i) the state vector $\mathbf{u}$ is a Markov process: $\pi\left(\mathbf{u}^{(k)}|\mathbf{u}^{(0:k-1)}\right) = \pi\left(\mathbf{u}^{(k)}|\mathbf{u}^{(k-1)}\right)$; (ii) observations $\mathbf{s}^{(k)}$ are independent of state history $\mathbf{u}^{(0:k-1)}$: $\pi\left(\mathbf{s}^{(k)}|\mathbf{u}^{(0:k)}\right) = \pi\left(\mathbf{s}^{(k)}|\mathbf{u}^{(k)}\right)$. The Bayes theorem can then be reformulated, for a time-dependent system, in the following form :

$$\pi\left(\mathbf{u}^{(k)}|\mathbf{s}^{(0:k)}\right) = \frac{\pi\left(\mathbf{u}^{(k)}|\mathbf{s}^{(0:k-1)}\right)\pi\left(\mathbf{s}^{(k)}|\mathbf{u}^{(k)}\right)}{\pi\left(\mathbf{s}^{(k)}|\mathbf{s}^{(0:k-1)}\right)} \tag{3}$$

In order to build the Kalman filter, we consider a linear dynamical system and it is necessary to add the three following hypotheses : (i) observation and model errors are statistically independent of the state; (ii) observation and model errors are statistically independent of each other; (iii) all probability density functions introduced in the Kalman filter are assumed to be Gaussian pdf. Moreover, pdf related to measurement and model errors are Gaussian white noise. Finally, introducing these hypotheses in the dynamical system (2) and using Bayes theorem (3) we can show that the Kalman filtering can be decomposed in two steps :

(i) a prediction step, wherein an *a priori* estimation $\mathbf{u}^{(k+\frac{1}{2})}$ of the state vector $\mathbf{u}^{(k+1)}$ is realized

(ii) a correction or assimilation step, wherein the *a priori* estimation is corrected by observed data $\mathbf{s}$ in order to obtain the *a posteriori* state vector $\mathbf{u}_a^{(k+1)}$ which is the Kalman filter estimation of the real state.

To solve inverse problems using Kalman filtering, it is necessary to explicitly introduce the parameters vector $\mathbf{p}$, which contains parameters we search to identify, in the dynamical system. We thus define an expanded state vector, noted $\bar{\mathbf{u}}$, and the associated dynamical system (4).

$$\begin{cases} \bar{\mathbf{u}}^{(k+1)} = \bar{\mathcal{M}}\bar{\mathbf{u}}^{(k)} + \bar{\mathbf{e}}_M^{(k)} \\ \mathbf{s}^{(k)} = \bar{\mathcal{H}}\bar{\mathbf{u}}^{(k)} + \mathbf{e}_S^{(k)} \end{cases} \quad \text{with } \bar{\mathbf{u}}^{(k)} = \begin{bmatrix} \mathbf{u}^{(k)} \\ \mathbf{p}^{(k)} \end{bmatrix} \tag{4}$$

The idea is then to apply a non-linear Kalman filter to the dynamical system (4), we consider here the Unscented Kalman Filter [8]. The main drawback of this approach, besides its low robustness with highly corrupted data, is the computational cost which is in $O((n + n_p)^3)$. The idea is then to use a similar approach as in [1] in order to circumvent these two drawbacks. More specifically, we introduce the modified Constitutive Relation Error in the dynamical system definition.

3.2. The modified Constitutive Relation Error

The modified Constitutive Relation Error (mCRE) is a deterministic method used to solve identification problems based on energy functional minimization [11, 7, 2]. In the context of thermal transient problems, the mCRE functionnal is defined as follows :

$$\mathcal{E}_m^2(u, \mathbf{q}; \mathbf{p}, s) = \frac{1}{2}\int_{I_t}\int_{\Omega} (\mathbf{q} - \mathcal{K}\nabla u)\, \mathcal{K}^{-1} (\mathbf{q} - \mathcal{K}\nabla u)\, d\mathbf{x}dt + \frac{1}{2}\frac{r}{1-r}\int_{I_t} \|\Pi u - \mathbf{s}\|^2 dt$$

where $u \in \mathcal{U}$ is the temperature field, $\mathbf{q} \in \mathcal{S}$ the thermal flux and $\mathcal{K}$ the thermal conductivity operator. $r \in [0, 1[$ is a scalar penalty coefficient used to weight more or less the measurement error term, classically it is used $r = 0.5$ even though an optimal value can be obtained related to the measurement noise (Morozov principle). We define the two admissibility spaces $\mathcal{U}(I_t) = \left\{u \in H^1(\Omega) \otimes L^2(I_t) \setminus u = u^d\, \forall (x,t) \in \partial\Omega_u \times I_t,\, u = u^0 \forall (\mathbf{x}, t) \in \Omega \times \{I_t(0)\}\right\}$ and $\mathcal{S}(I_t) = \left\{\mathbf{q} \in L^2(\Omega) \otimes L^2(I_t) \setminus \mathbf{q} = \mathbf{q}^d\, \forall (\mathbf{x}, t) \in \partial\Omega_q \times I_t,\, c\dot{u} + \nabla \cdot \mathbf{q} = f\, \forall (\mathbf{x}, t) \in \Omega \times I_t\right\}$ where $I_t(0)$ denotes the first value of time interval I_t.

The classical way to solve inverse problems using mCRE is then to define $\mathbf{p} \in \mathcal{P}$, where $\mathcal{P}$ is the parameters admissibility space, as the solution of (5).

$$\mathbf{p} = \underset{\xi \in \mathcal{P}}{\text{argmin}} \min_{(u,\mathbf{q}) \in \mathcal{U} \times \mathcal{S}} \mathcal{E}_m^2(u, \mathbf{q}; \xi, s) \tag{5}$$

In practice problem (5) is solved using an alternate minimization associated to a fixed point algorithm :

M1 minimization over $\mathcal{U} \times \mathcal{S}$: $(u_{ad}, \mathbf{q}_{ad}) = \underset{(u,\mathbf{q}) \in \mathcal{U} \times \mathcal{S}}{\text{argmin}}\ \mathcal{E}_m^2(u, \mathbf{q}; \mathbf{p}, \mathbf{s})$

M2 minimization over $\mathcal{P}$: $\mathbf{p} = \underset{\boldsymbol{\xi}\in\mathcal{P}}{\operatorname{argmin}}\, \mathcal{E}^2_m(u_{ad}, \mathbf{q}_{ad}; \boldsymbol{\xi}, \mathbf{s})$

Numerically the most expensive step is the first minimization since it is necessary to solve a coupled forward-backward problem in time. Therefore, if we use a monolithic scheme like linear interpolation in time, it is necessary to solve a $2 \times n \times n_t$ linear system.

3.3. Coupling between the Kalman filter and the mCRE

The approach we propose here is to introduce the modified Constitutive Relation Error, and precisely the associated admissible fields computation, in the dynamical system (2). We can then write a new dynamical system as :

$$\begin{cases} \mathbf{p}^{(k+1)} = \mathbf{p}^{(k)} + \mathbf{e}_P^{(k)} \\ \mathbf{s}^{(k)} = \mathcal{H}_m\left(\mathbf{p}^{(k)}, \mathbf{s}^{(k-1:k)}\right) + \mathbf{e}_S^{(k)} \end{cases} \tag{6}$$

where the first equation translates the fact that without *a priori* knowledge about parameters time dependence we add a stationarity hypothesis and we perturb it by a Gaussian white noise. The new observation operator is then defined as follows :

$$\mathcal{H}_m\left(\mathbf{p}^{(k)}, \mathbf{s}^{(k-1:k)}\right) := \mathcal{H} \circ \mathrm{mCRE}_1\left(\mathbf{p}^{(k)}, \mathbf{s}^{(k-1:k)}\right)$$

with $\mathbf{s}^{(k-1:k)}$ used to denote $(\mathbf{s}^{(k-1)}, \mathbf{s}^{(k)})$. The application $\mathrm{mCRE}_1(\mathbf{p}^{(k)}, \mathbf{s}^{(k-1:k)})$ returns the field $u_{ad}^{(k)}$ resulting from minimization **M1**. A small difference comes from the fact that in a classical way minimization **M1** is over $\mathcal{U}(I_t) \times \mathcal{S}(I_t)$ however here we define admissible fields for each subinterval $I_t^{(k)} = [t^{(k-1)}, t^{(k)}]$ so that we realize minimization according to $\mathcal{U}(I_t^{(k)}) \times \mathcal{S}(I_t^{(k)})$. In order to solve the identification problem we apply the Unscented Kalman Filter to dynamical system (6), we denote this approach Modified Kalman Filter (MKF). Advantages to reformulate the dynamical system under form (6) are of three kinds :

(i) The fact to introduce the modified Constitutive Relation Error in the Kalman filter increases its robustness compared to the classical formulation;
(ii) With this new formulation, the Kalman filter computation cost is considerably reduced as we pass from $O((n+n_p)^3)$ to $O(n_p^3)$. However we introduce a new computation cost source with the resolution of the first mCRE minimization;
(iii) The state system computation being based on mCRE admissible fields computation, it is possible to introduce reduced order modelling based on Proper Generalized Decomposition.

3.4. Reduced order modelling

In order to reduce the computation cost and speed-up identification process, we introduce reduced order modelling based on Proper Generalized Decomposition (PGD) [5]. The PGD basic idea is to compute in an *offline* step the solution under a separated variable representation and evaluate this solution in an *online* step. Another advantage to use PGD comes from the fact that is possible to include several parameters as extra-coordinates. The identification strategy we propose being an incremental method, the system state over each time increment depends on parameters $\mathbf{p}^{(k)}$, observations data $(\mathbf{s}^{(k-1)}, \mathbf{s}^{(k)})$ and initial condition $u_0^{(k)} = u^{(k-1)}$. The main difficulty is associated to the initial condition which is non-homogeneous. To circumvent this problem we choose to project it on a reduced basis from a first PGD decomposition, we then write $u_0^{(k)} = \sum_{i=1}^{n_{init}} \gamma_i \psi_i(\mathbf{x})$ and we introduce only the γ_i coefficients as extra-coordinates. Consequently, we construct a PGD approximation on the following form :

$$u \simeq \sum_{i=1}^{M} \prod_{j=1}^{n_{init}} \Gamma_i^{(j)}(\gamma_i) \cdot \prod_{l=1}^{m} \theta_i^{(l)}(s_l^{(k-1)}) \prod_{q=1}^{m} \beta_i^{(q)}(s_l^{(k)}) \cdot \phi_i(\mathbf{p}) \cdot \lambda_i(t) \cdot w_i(\mathbf{x})$$

4. Numerical results

The strategy we propose here is applied on two parabolic problems (transient thermal problems) and results are compared to those obtained using the classical approach based on Unscented Kalman Filter (UKF). In order to simulate observations data the strategy used is the following :

(i) Solve the direct problem with true parameters values using the same space mesh than the one used for inverse problem but considering fine time discretization;

(ii) Extract temperature nodal values according to sensor placement and acquisition time (coarser than for direct resolution);

(iii) Add a Gaussian white noise : $s_i^{(k)} = u_i^{(k)}\,(1 + \mathcal{N}(0, \Sigma))\ \ \forall i = 1, ..., m\ ;\ \forall k$

The first example we consider is the identification of time-dependent Neumann boundary condition. The problem setting is given in Figure 1(a) and we consider highly corrupted data with $\Sigma = 0.2$, that is to say 20% measurement noise. The identification results are presented in Figures (1(b),1(c)).

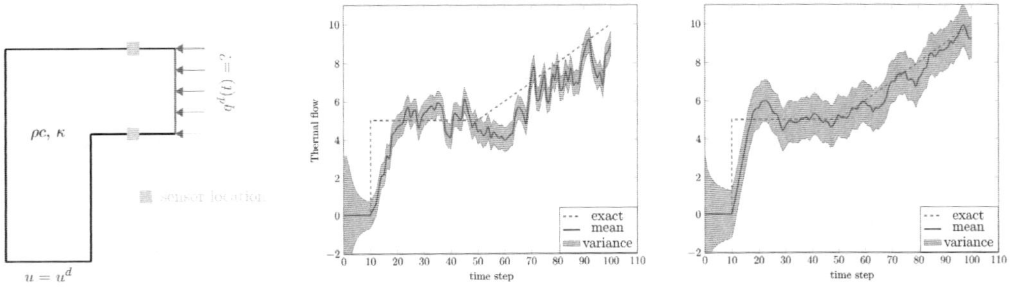

(a) Problem setting (b) UKF identification (c) MKF identification

Figure 1. Thermal-flow identification using UKF and MKF approaches with 20% measurement noise

In order to study tuning parameters influence on results obtained by the Modified Kalman Filter algorithm we realize a parametric study. We show in Figure 2(a) the influence of variance associated to parameters error C_P and observation error variance C_S in term of global error $\varepsilon_{MKF} = \frac{\|\mathbf{p}_{true} - \mathbf{E}[\mathbf{p}_{MKF}]\|_{L^2(I_t)}}{\|\mathbf{p}_{true}\|_{L^2(I_t)}}$. And in Figure 2(b) we show the influence of mCRE penalty coefficient r and measure noise level Σ.

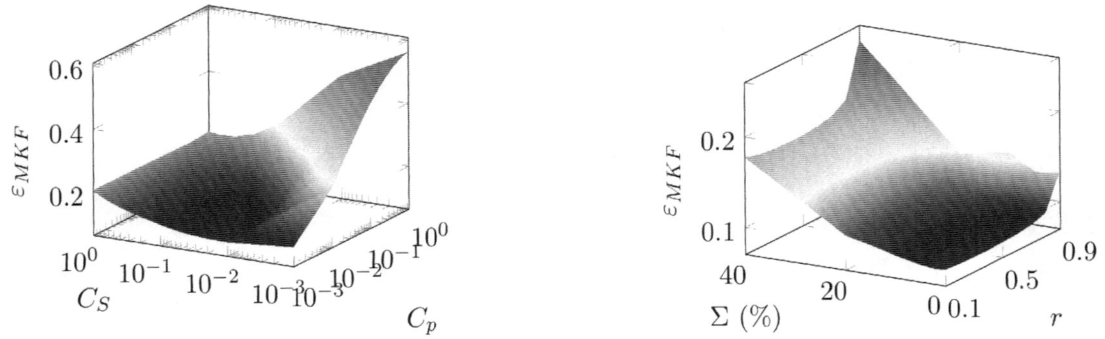

(a) C_p and C_S influence (b) Penalization and measure noise influence

Figure 2. Modified Kalman filter parametric study

The second problem treated with the MKF approach is the identification of time-independent material parameters in a multi-layers structure. The problem setting is given in Figure 3(a) and we consider a measurement noise of 10%, $\Sigma = 0.1$. The results are shown in Figures (3(b),3(c)) only for κ_1 identification but the identification of other parameters performs the same way.

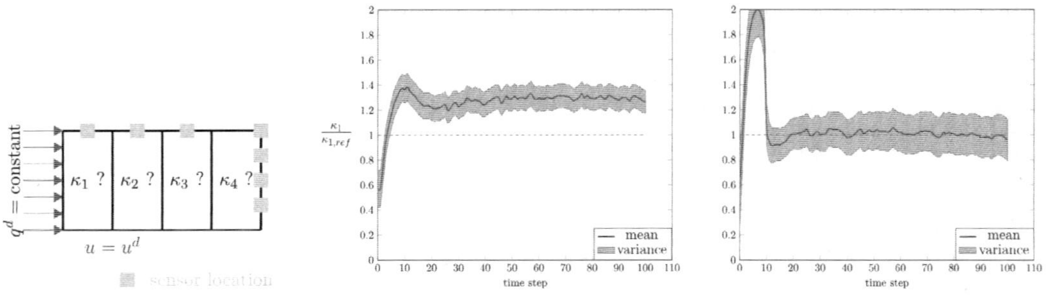

(a) Problem setting (b) κ_1 UKF identification (c) κ_1 MKF identification

Figure 3. Thermal conductivity identification using UKF and MKF approaches with 10% measurement noise

To conclude, using the MKF approach we observe : (i) a significant computation time gain; (ii) a better response of the identification process to abrupt changes of parameters to be identified; (iii) global better results on parameters identification.

References

[1] A. Alarcon, C. Bodel, M. Bonnet. *A coupled unscented Kalman filter and modified error in constitutive relation technique for structural dynamics identification*, Proc. EURODYN, 2163-2168, 2011.

[2] O. Allix, P. Feissel, H. Nguyen. *Identification strategy in the presence of corrupted measurements*, Engineering Computations, 487-504, 2005.

[3] G. Bolzon, R. Fedele, G. Maier. *Parameter identification of a cohesive crack model by Kalman filter*, Computer Methods in Applied Mechanics and Engineering, 2847-2871, 2002.

[4] R. Bouclier, F. Louf, L. Chamoin. *Real-time validation of mechanical models coupling PGD and constitutive relation error*, Computationnal Mechanics, 861-883, 2013.

[5] F. Chinesta, R. Keunings, A. Leygue. *The Proper Generalized Decomposition for advanced Numerical Simulations*, Springer, 2014.

[6] F. Darema. *Dynamic Data Driven Applications Systems : A new paradigm for applications simulations and measurements*, Computational Science-ICCS 2004: 4th International Conference, Springer, 662-669, 2004.

[7] A. Deraemaeker, P. Ladevèze. *A constitutive relation error model-updating method in structural dynamics with uncertain measurements*, Inverse Problems in Engineering Mechanics, 273-283, 2003.

[8] S. Julier, J. Uhlmann., *A new extension of the kalman filter to nonlinear systems*, The 11th Int. Symposium on Aerospace/Defense Sensing and Simulation and Controls, 1996.

[9] J. Kaipio, E. Somersalo. *Statistical and Computational Inverse Problems*, Springer, 2006.

[10] R. Kalman. *A new approach to linear filtering and prediction problems*, Journal of Basic Engineering, 35-45, 1960.

[11] P. Ladevèze, D. Nedjar, M. Reynier. *Updating of finite element models using vibration tests*, AIAA Journal, 1485-1491, 1994.

[12] H. Nguyen, O. Allix, P. Feissel. *A robust identification strategy for rate-dependent models in dynamics*, Inverse Problems, 2008.

[13] E. Prudencio, P. Bauman, D. Faghihi, K. Ravi-Chandar, J. Oden. *A computational framework for dynamic data driven material damage control, based on bayesian inference and model selection*, International Journal for Numerical Methods in Engineering, 2014

Journal of Physics: Conference Series **657** (2015) 012008 doi:10.1088/1742-6596/657/1/012008

Nonlinear electromagnetic inverse scattering in via Frozen or Broyden update of the Fréchet derivative

E. Tavanti [1], **C. Estatico** [2], **A. Fedeli** [1], **M. Pastorino** [1], **A. Randazzo** [1]

[1]Department of Electrical, Electronic, Telecommunication Engineering, and Naval Architecture, University of Genoa, Via Opera Pia 11A, I-16145, Genoa, Italy.

[2]Department of Mathematics, University of Genoa, Via Dodecaneso 35, I-16146 Genoa, Italy.

E-mail: emanuele.tavanti@gmail.com

Abstract. Microwave imaging methods are useful for non-destructive inspection of dielectric targets. In this work, a numerical technique for solving the 3D Lippmann-Schwinger integral equation of the inverse scattering problem via Gauss-Newton linearization in Banach spaces is analysed. More specifically, two different approximations of the Fréchet derivative are proposed in order to speed up the computation. Indeed it is well known that the computation of the Fréchet derivative is generally quite expensive in three dimensional restorations. Numerical tests show that the approximations give a faster restoration without loosing accuracy.

1. Introduction

As it is well known the electromagnetic inverse scattering problem is the basic formulation for microwave imaging methods. Although this problem has been suitably studied from a theoretical point of view, new methods are continuously developed to face new and challenging applications. To inspect dielectric targets, several different techniques can be adopted. These techniques has been devised in the last decades starting from some pioneering research results [1],[2]. Some examples can be found in [3]-[12] and in the references therein.

In this context, an approach for solving the Lippmann-Schwinger integral equation of the inverse scattering problem by lossy dielectric targets has been proposed in [13]. It is based on an inexact-Newton method and constituted by two nested loops, and it has been tested versus experimental input data [14]. A review of the obtained results have been reported in [15]. Although such results appear to be quite satisfactory, in particular in dealing with the ill-posedness of the considered inverse problem, one of the main drawback is related to the significant over-smoothing effects associated to the properties of the Hilbert space in which the approach has been developed. In order to mitigate this effect, a new formulation has been proposed in [16], in which the inverse scattering equations have been solved in L^p Banach spaces.

Despite the increasing of the formulation complexity, the new approach has led to a noticeable reduction in the reconstruction errors as compared with solutions developed in the Hilbert spaces. So for, the new approach has been applied to two-dimensional configurations for which a scalar formulation can be used. In the present paper, this method is preliminary extended to treat three-dimensional targets, where the computation of the Fréchet derivative becomes numerically expensive. This way, two different low-cost approximations of the Fréchet derivative, namely the frozen [17] and the Broyden update [18], are investigated in order to obtain a faster reconstruction algorithm.

---

[1] E-mail: emanuele.tavanti@gmail.com ; alessandro.fedeli@edu.unige.it; matteo.pastorino@unige.it; andrea.randazzo@unige.it.

[2] E-mail: estatico@dima.unige.it. The work of C. Estatico is partly supported by PRIN 2012 N. 2012MTE38N.

2. The 3D microwave inverse scattering model

An unknown dielectric target, located inside an investigation cube V_{inv}, is illuminated by one or more incident known electric fields, generated by proper emitting antennas located outside the cube. For any illumination, at any point $\mathbf{r}$ of the whole three-dimensional space, the electromagnetic interaction between the dielectric object and the incident radiation gives rise to a total electric field $\mathbf{E}_{tot}(\mathbf{r}) = \mathbf{E}_{inc}(\mathbf{r}) + \mathbf{E}_{scatt}(\mathbf{r})$, which is the sum of the incident and the scattered electric fields.

In the inverse scattering model, the measurement of $\mathbf{E}_{tot}(\mathbf{r})$ by several receiving antennas located in a measurement domain V_{obs} external to the investigation cube allows to reconstruct (a suitable approximation of) the unknown dielectric object. In our numerical simulations, the emitting and receiving antennas are uniformly distributed on a fixed sphere enclosing the investigation cube. For sake of simplicity, a single-view case is described in the following. The extension of the mathematical formulation to the multi-view case is however straightforward.

The scattered electric field $\mathbf{E}_{scatt}(\mathbf{r})$ is related to the dielectric properties of the investigated area V_{inv} by means of the following two integral equations, usually referred as data and state equations,

$$\begin{aligned} \mathbf{E}_{scatt}(\mathbf{r}) &= \mathbf{G}_{data}(c\mathbf{E}_{tot})(\mathbf{r}), & \mathbf{r} \in V_{obs} \\ \mathbf{E}_{inc}(\mathbf{r}) &= \mathbf{E}_{tot}(\mathbf{r}) - \mathbf{G}_{state}(c\mathbf{E}_{tot})(\mathbf{r}), & \mathbf{r} \in V_{inv} \end{aligned} \tag{1}$$

where $c(\mathbf{r}) = \epsilon_r(\mathbf{r}) - 1$ is the target contrast function (being ϵ_r the space dependent relative complex dielectric permittivity of the cube V_{inv}), the linear convolution operators $\mathbf{G}_{data}$ and $\mathbf{G}_{state}$ are

$$\mathbf{G}_{data}\mathbf{f}(\mathbf{r}) = -k_0^2 \int_{V_{inv}} \mathbf{f}(\mathbf{r}) \cdot \bar{\mathbf{G}}_0(\mathbf{r},\mathbf{r}')d\mathbf{r}', \qquad \mathbf{r} \in V_{obs},$$

$$\mathbf{G}_{state}\mathbf{f}(\mathbf{r}) = -k_0^2 \int_{V_{inv}} \mathbf{f}(\mathbf{r}) \cdot \bar{\mathbf{G}}_0(\mathbf{r},\mathbf{r}')d\mathbf{r}', \qquad \mathbf{r} \in V_{inv},$$

being $k_0 = \omega\sqrt{\epsilon_0\mu_0}$ the free-space wavenumber and $\bar{\mathbf{G}}_0(\mathbf{r},\mathbf{r}') = -\frac{1}{4\pi}\left(\bar{\mathbf{I}} + \frac{\nabla\nabla}{k_0^2}\right)\frac{e^{-jk|\mathbf{r}-\mathbf{r}'|}}{|\mathbf{r}-\mathbf{r}'|}$ is the free-space dyadic Green's function. The data and state equations can be merged in order to obtain the following single non-linear operator equation

$$\mathbf{E}_{scatt}(\mathbf{r}) = \mathbf{G}_{data}\big(\mathbf{C}((\mathbf{I} - \mathbf{G}_{state}\mathbf{C})^{-1}\mathbf{E}_{inc})\big)(\mathbf{r}), \tag{2}$$

where $\mathbf{r} \in V_{obs}$ and $\mathbf{C}$ is the operator such that $\mathbf{C}(\mathbf{f})(\mathbf{r}) = c(\mathbf{r})\mathbf{f}(\mathbf{r})$, $\mathbf{r} \in V_{inv}$. By defining the non-linear operator $\mathbf{F}(c) = \mathbf{G}_{data}\big(\mathbf{C}((\mathbf{I} - \mathbf{G}_{state}\mathbf{C})^{-1}\mathbf{E}_{inc})\big)$ that maps the contrast function c with the scattered electric field $\mathbf{E}_{scatt}$, equation (2) can be written in compact form as follows

$$\mathbf{F}(c)(\mathbf{r}) = \mathbf{E}_{scatt}(\mathbf{r}), \qquad \mathbf{r} \in V_{obs}. \tag{3}$$

The non-linear equation (3) models the full inverse scattering problem: Given the scattered electric field $\mathbf{E}_{scatt}$ (i.e., the data), measured in the measurement domain V_{obs}, find the contrast function c (i.e., the unknown) in the investigation domain V_{inv} such that $\mathbf{F}(c) = \mathbf{E}_{scatt}$.

3. The solution of the inverse problem in Banach spaces via quasi-Newton schemes

By a pure mathematical point of view, the nonlinear operator of (3) is a map $\mathbf{F}: C \to E$ between a linear functional space C of the contrast functions and the functional space E of the electric fields. In the classical solving schemes, C and E are $L^\infty(V_{inv}) \subset L^2(V_{inv})$ and $L^2(V_{obs})$ respectively, being L^2 the Hilbert spaces of square-integrable functions. In our solving scheme, C and E are the more general Banach spaces $L^\infty(V_{inv}) \subset L^p(V_{inv})$ and $L^p(V_{obs})$ of p-integrable functions, with $1 < p < 2$, so that the more recent theory of regularization in Banach spaces can be applied [19]-[21]. Indeed, it has been shown that the resolution of functional equations in L^p Banach spaces, with $1 \le p < 2$, reduces over-smoothness effects of the classical L^2 solutions, leading to a better localization of the discontinuities of the scatterers.

To solve (3), a Newton-like iterative method is applied to minimize the $L^p(V_{obs})$ residual functional

$\Omega_p: L^p(V_{inv}) \to \mathbb{R}$ defined as follows

$$\Omega_p(c) = \frac{1}{p}\|\mathbf{F}(c) - \mathbf{E}_{scatt}\|^p_{L^p(V_{obs})}.$$

In particular, any linearized equation of the Newton scheme is solved by means of a gradient-like iterative regularization method for inverse problems in Banach spaces [19]. In our three-dimensional case, the Fréchet derivative $\mathbf{F}'_c: L^p(V_{inv}) \to L^p(V_{obs})$ of the operator $\mathbf{F}$ at point c is given by

$$\mathbf{F}'_c\delta(\mathbf{r}) = \mathbf{G}^c_{data}(\delta\mathbf{E}^c_{tot})(\mathbf{r}),\ \mathbf{r} \in V_{obs},\quad \forall\delta \in L^p(V_{inv}), \tag{4}$$

where $\mathbf{E}^c_{tot}(\mathbf{r}) = ((\mathbf{I} - \mathbf{G}_{state}\mathbf{C})^{-1}\mathbf{E}_{inc})(\mathbf{r}),\ \mathbf{r} \in V_{inv}$, and the operator $\mathbf{G}^c_{data}$ is given by

$$\mathbf{G}^c_{data}\mathbf{f}(\mathbf{r}) = -k_0^2\int_{V_{inv}} \mathbf{f}(\mathbf{r}) \cdot \overline{\mathbf{G}}_c(\mathbf{r},\mathbf{r}')d\mathbf{r}',\ \mathbf{r} \in V_{obs}, \tag{5}$$

being $\overline{\mathbf{G}}_c$ the dyadic Green's function for an inhomogeneous background characterized by c [22].

As it is well-known, together to convergence issues, the drawback of Newton methods in real applications is the generally expensive computation of the new Fréchet derivative at each iteration. On the other hand, it is expected that the "new" Fréchet derivative at iteration k could be suitably approximated by means of the "previous" one at iteration $k-1$. Basically, this is the key point of several generalizations named as *quasi-Newton* methods. In this paper we deal with the frozen-Newton method [17] and the Broyden-Newton method [18], as summarized in the following steps.

I) Let $c_0 \in L^\infty(V_{inv})$ be an initial guess (the null function c_0 is often used in the applications), and comupute the Fréchet derivative of $\mathbf{F}$ at point c_0, denoted as $\mathbf{F}'_0$. Set $n := 0$ (please notice that n denotes the outer interation index).

II) Frozen update: if $n \bmod s = 0$ then $\tilde{\mathbf{F}}'_n = \mathbf{F}'_n$, that is, let $\tilde{\mathbf{F}}'_n$ be the (new) Fréchet derivative $\mathbf{F}'_n$ of $\mathbf{F}$ at point c_n; else set $\tilde{\mathbf{F}}'_n = \tilde{\mathbf{F}}'_{n-1}$, (i.e., use the same derivative of the previous step)
Broyden update: if $n \bmod s = 0$ then $\tilde{\mathbf{F}}'_n = \mathbf{F}'_n$, else let $\tilde{\mathbf{F}}'_n$ be the approximation of the Fréchet derivative $\mathbf{F}'_n$ of $\mathbf{F}$ at point c_n defined by the following Broyden update

$$\tilde{\mathbf{F}}'_n = \tilde{\mathbf{F}}'_{n-1} + \frac{(\mathbf{F}(c_n) - \mathbf{F}(c_{n-1})) - \tilde{\mathbf{F}}'_{n-1}\delta_{n-1}}{\|\delta_{n-1}\|^2_{L^2(V_{inv})}}\delta^T_{n-1}.$$

III) Compute a regularized solution of the linear system in $L^p(V_{inv})$ with respect to the unknown function δ_n

$$\tilde{\mathbf{F}}'_n\delta_n = \mathbf{E}_{scatt} - \mathbf{F}(c_n) \tag{6}$$

by means of the minimization of the functional $\frac{1}{p}\left\|\tilde{\mathbf{F}}'_n\delta_n - (\mathbf{E}_{scatt} - \mathbf{F}(c_n))\right\|^p_{L^p(V_{obs})}$ in $L^p(V_{inv})$ Banach space, for $1 < p < 2$. To this end, we consider an early stop of the following Landweber iterative method for linear equations in $L^p(V_{inv})$ [23].
Let $\delta^*_{n,0} = 0 \in L^q(V_{inv})$ be the inner initial guess. Then for $k = 0,1,2,\ldots$ (please notice that k denotes the inner interation index), compute the iterative steps

$$\delta^*_{n,k+1} = \delta^*_{n,k} - \tau_k\tilde{\mathbf{F}}'^*_n\mathbf{J}_p\left(\tilde{\mathbf{F}}'_n\delta_{n,k} - (\mathbf{E}_{scatt} - \mathbf{F}(c_n))\right), \tag{7}$$

$$\delta_{n,k+1} = \mathbf{J}_q(\delta^*_{n,k+1}), \tag{8}$$

where q is the Hölder conjugate of p, that is, $\frac{1}{p} + \frac{1}{q^*} = 1$, until a stopping rule is satisfied (e.g., a maximum number of inner iterations k_{max} is reached or the norm of the functional to be minimized falls below a fixed threshold). Here $\tau_k > 0$ is the suitable step size, $\mathbf{J}_p: L^p(V_{obs}) \to L^q(V_{obs})$ and $\mathbf{J}_q: L^q(V_{inv}) \to L^p(V_{inv})$ denote the normalized duality maps between Banach spaces defined as

$\mathbf{J}_s(v) = \partial\left(\frac{1}{r}\|v\|_{L^s}^2\right) = \|v\|_{L^s}^{2-s}|v|^{s-1}e^{j\arg(v)}$, if $v \neq 0$, and $\mathbf{J}_s(0) = 0$, for any $s > 1$ [20]. Set $\delta_n := \delta_{n,k_{max}}$.

IV) Update the current quasi-Newton solution by setting $c_{n+1} = c_n + \delta_n$ and let $n := n + 1$. Stop if a predefined stopping rule (e.g., the discrepancy principle $\Omega_p(c_n) < p\epsilon^p$, being ϵ an estimate of the $L^p(V_{obs})$-norm of the noise on $\mathbf{E}_{scatt}$) is satisfied elsewhere go to step II.

A. *The Frozen and the Broyden updates*

In the frozen method, the same Fréchet derivative computed at iteration k is used (i.e, is "frozen") for a fixed number s of subsequent iterations $k+1, \dots, k+s-1$ [17].

In the Broyden method, a special low-cost (i.e., rank-1) update of the Fréchet derivative is computed at each iteration [18], as now sketched. Differing from the straightforward one dimensional case, the difference $\mathbf{F}(c_n) - \mathbf{F}(c_{n-1}) \in L^p(V_{obs})$ does not give us enough information to approximate the whole linear operator $\mathbf{F}'_{c_n}$. On the other hand, the difference $\mathbf{F}(c_n) - \mathbf{F}(c_{n-1})$ gives us information about the directional derivative of $\mathbf{F}$ on the direction $\delta_{n-1} = c_n - c_{n-1}$, since $\mathbf{F}'_{c_n}\delta_{n-1} = \mathbf{F}(c_n) - \mathbf{F}(c_{n-1}) + O(\|\delta_{n-1}\|^2)$. On this ground, the Broyden update modifies the Fréchet derivative along the direction $\delta_{n-1} = c_n - c_{n-1}$ only, by means of the increment $\mathbf{F}(c_n) - \mathbf{F}(c_{n-1})$, leaving unchanged the operator on the subspace orthogonal to δ_{n-1}. With standard Hilbertian decomposition $v = \alpha\delta_{n-1} + u$, where $u \perp v$, $\alpha = \frac{\delta_{n-1}^T}{\|\delta_{n-1}\|^2} v$ and $\|u\| = 1$, we have that

$$\mathbf{F}'_{c_n}v = \mathbf{F}'_{c_n}(\alpha\delta_{n-1} + u) = \alpha\mathbf{F}'_{c_n}\delta_{n-1} + \mathbf{F}'_{c_n}u = \alpha\mathbf{F}'_{c_n}\delta_{n-1} - \alpha\mathbf{F}'_{c_{n-1}}\delta_{n-1} + \alpha\mathbf{F}'_{c_{n-1}}\delta_{n-1} + \mathbf{F}'_{c_n}u \approx \alpha\mathbf{F}'_{c_n}\delta_{n-1} - \alpha\mathbf{F}'_{c_{n-1}}\delta_{n-1} + \alpha\mathbf{F}'_{c_{n-1}}\delta_{n-1} + \mathbf{F}'_{c_{n-1}}u = \alpha\big(\mathbf{F}(c_n) - \mathbf{F}(c_{n-1})\big) - \alpha\mathbf{F}'_{c_{n-1}}\delta_{n-1} + \mathbf{F}'_{c_{n-1}}v = \mathbf{F}'_{c_{n-1}}v + \left[\frac{1}{\|\delta_{n-1}\|^2}\left(\mathbf{F}(c_n) - \mathbf{F}(c_{n-1}) - \mathbf{F}'_{c_{n-1}}\delta_{n-1}\right)\delta_{n-1}^T\right]v,$$

that is,

$$\mathbf{F}'_{c_n} \approx \mathbf{F}'_{c_{n-1}} + \frac{\mathbf{F}(c_n) - \mathbf{F}(c_{n-1}) - \mathbf{F}'_{c_{n-1}}\delta_{n-1}}{\|\delta_{n-1}\|^2}\delta_{n-1}^T .$$

4. A numerical validation

The data and state scattering equations (1) are discretized by using pulse basis functions and Dirac's delta weighting functions [24]. In particular, the investigation domain V_{inv} is discretized into N cubic voxels of centers $\mathbf{r}_n^{inv}$, $n = 1, \dots, N$, and the observation domain is composed by M measurement points located at positions $\mathbf{r}_m^{meas}$, $m = 1, \dots, M$. We focus on the investigation domain shown in Figure 1; it contains a cube and a parallelepiped of equal relative dielectric permittivity and it is partitioned with $N = 10648$. The measurement points are $M = 50$, uniformly located on a sphere of radius $2\,m$ and centered in $(0;0;0)$. Six views are used, with plane waves at $300\,MHz$ as illuminating sources. It is worth noting that, in the numerical implementation of the approach, a BiCGStab-FFT algorithm [24] has been used to speed up the computation of the total internal field and of the inhomogeneous dyadic Green's function [22]. To achieve a comparison, we use $p \in \{1.2, 1.3, \dots, 2.2\}$ inside the previously introduced updating methods for Fréchet derivatives, with $s = 4$, and the complete one. We set maximum numbers of outer and inner iterations equal to 20 and 10, respectively, without stopping thresholds.

The following errors are used to quantify the reconstruction quality of the objects and background

$$E_{obj} = \frac{1}{N_{obj}}\sum_{\mathbf{r}_i^{inv} \in V_{obj}} \frac{\left|\hat{c}\left(\mathbf{r}_i^{inv}\right) - c\left(\mathbf{r}_i^{inv}\right)\right|}{\left|c\left(\mathbf{r}_i^{inv}\right) + 1\right|} \quad E_b = \frac{1}{N_b}\sum_{\mathbf{r}_i^{inv} \in V_b} \frac{\left|\hat{c}\left(\mathbf{r}_i^{inv}\right) - c\left(\mathbf{r}_i^{inv}\right)\right|}{\left|c\left(\mathbf{r}_i^{inv}\right) + 1\right|}.$$

where V_{obj} is objects' domain (N_{obj} voxels), V_b is background's one (N_b) and $\hat{c}\left(\mathbf{r}_i^{inv}\right)$ are contrast function's reconstructed values. Table 1 summarizes the results.

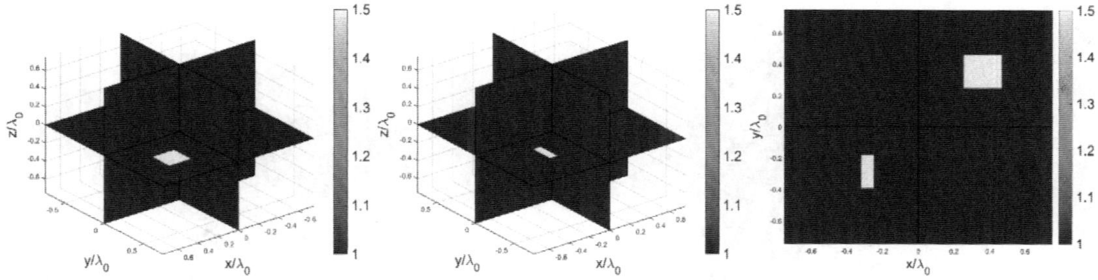

Figure 1. Actual distribution of the relative dielectric permittivity. Three-dimensional views (with slices along the planes $x = 0$, $y = 0$ and $z = 0$) and two-dimensional cut on the x-y plane.

p	Frozen			Broyden			Complete		
	E_{obj}	E_b	Time [s]	E_{obj}	E_b	Time [s]	E_{obj}	E_b	Time [s]
1.2	0.20	$1.5\ 10^{-3}$	531	0.23	$0.9\ 10^{-3}$	427	0.20	$1.6\ 10^{-3}$	2816
1.4	0.25	$2.6\ 10^{-3}$	366	0.27	$3\ 10^{-3}$	378	0.25	$2.6\ 10^{-3}$	1820
1.6	0.27	$3.4\ 10^{-3}$	359	0.29	$3.6\ 10^{-3}$	373	0.27	$3.4\ 10^{-3}$	1794
1.8	0.29	$4.2\ 10^{-3}$	366	0.29	$4.4\ 10^{-3}$	369	0.29	$4.2\ 10^{-3}$	1800
2.0	0.29	$4.8\ 10^{-3}$	357	0.30	$4.8\ 10^{-3}$	372	0.29	$4.8\ 10^{-3}$	1790
2.2	0.30	$5.4\ 10^{-3}$	390	0.30	$5\ 10^{-3}$	385	0.30	$5.4\ 10^{-3}$	1821

Table 1. Errors and computational times upon p.

We see that as p approaches to 2.0 or higher values, both E_{obj} and E_b grow; this supports a choice of $1 < p < 2$ when we are interested in detection of relatively small discontinuities like those in our investigation domain. Figures 2-3 report the reconstructed dielectric scenes. Concerning the approximations of Fréchet derivatives, we note that both Broyden and frozen strategies drastically reduce the computational time by keeping errors very close to the exact derivative case.

5. Conclusions

In this paper, an approach for solving microwave imaging of lossy dielectric targets, which was originally developed for treating two-dimensional configurations, has been extended to inspecting three-dimensional targets. The approach is based on the vector integral equations of the inverse scattering problem (Green's tensor formulation) and on the application of an inexact-Newton method developed in L^p Banach spaces with approximation of the Fréchet derivatives. Some preliminary numerical results have been reported. Further activity will be devoted to inspecting more complex targets and to define rules for a proper choice of the optimum p-parameter, which is of course application-dependent, even by using real experimental data.

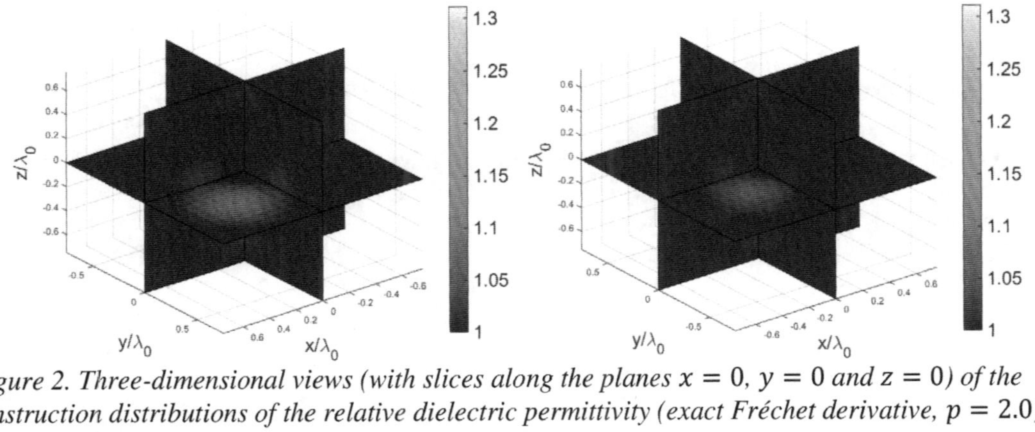

Figure 2. Three-dimensional views (with slices along the planes $x = 0$, $y = 0$ and $z = 0$) of the reconstruction distributions of the relative dielectric permittivity (exact Fréchet derivative, $p = 2.0$).

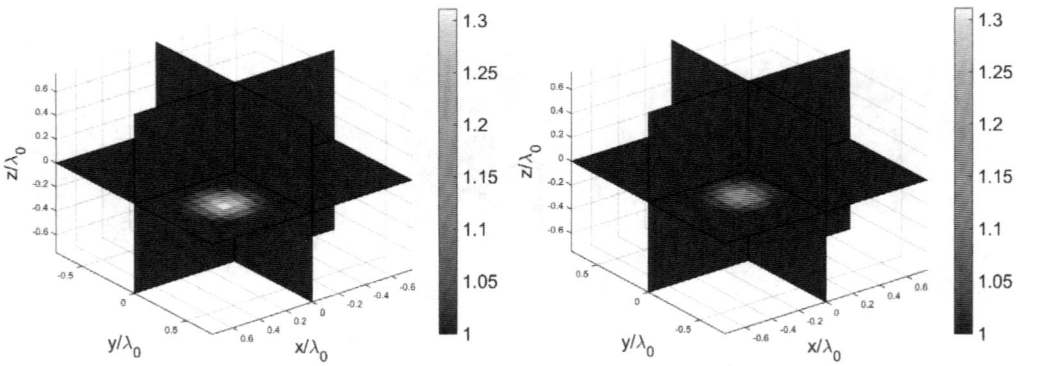

Figure 3. Three-dimensional views (with slices along the planes $x = 0$, $y = 0$ and $z = 0$) of the reconstruction distributions of the relative dielectric permittivity (exact Fréchet derivative, $p = 1.2$).

References

[1] Bolomey J C, Lesselier D, Pichot C and Tabbara W 1981 *IEEE Trans. Antennas Propagat.* **29** 206-212.
[2] Bolomey J C, Izadnegahdar A, Jofre L, Pichot C, Peronnet G and Solaimani M 1982 *IEEE Trans. Microwave Theory Tech.* **30** 1998-2000.
[3] Lesselier D and Bowler J 2002 Special issue on electromagnetic and ultrasonic nondestructive evaluation *Inv. Probl.* **18**.
[4] Chen C C, Johnson J T, Sato M and Yarovoy A G 2007 Special issue on subsurface sensing using ground-penetrating radar *IEEE Trans. Geosci. Remote Sens.* **45**.
[5] Zhang Z Q and Liu Q H 2004 *IEEE Trans. Biomed. Eng.* **51** 544-548.
[6] Gdoura S and Lesselier D 2012 *Proc. 2nd Int. Workshop on New Comput. Meth. for Inv.Probl.*
[7] van den Berg P M and Kleinman R E 1997 *Inv. Probl.* **13**. 1607-1620.
[8] Abubakar A, van den Berg P M and Mallorqui J J 2001 *IEEE Trans. Microwave Theory Tech.* **50** 1761-1771.
[9] De Zaeytijd J, Franchois A, Eyraud C and Geffrin J-M *IEEE Trans. Antennas Propagat.* **55** 3279-3292.
[10] Rubæk T, Meaney P M, Meincke P and Paulsen K D 2007 *IEEE Trans. Antennas Propagat.* **55** 2320-2331.
[11] Lambert M and Lesselier D 2000 *Inv. Probl.* **16** 563-576.
[12] Kooij B J, Lambert M and Lesselier D 1999 *Radio Sci.* **34** 1361-1371.
[13] Estatico C, Pastorino M and Randazzo A 2005 *IEEE Trans. Geosci Remote Sens.* **43** 2593-2505.
[14] Estatico C, Bozza G, Massa A, Pastorino M and Randazzo A 2005 *Inv. Probl.* **21** S81-94.
[15] Pastorino M 2010 *Microwave Imaging* (Hoboken: John Wiley & Sons).
[16] Estatico C, Pastorino M and Randazzo A 2012 *IEEE Trans. Antennas Propagat.* **60** 3373-3381.
[17] Jin Q 2010 *Mathematics of Computation* **79** 2191-2211.
[18] Broyden C G 1965 *Mathematics of Computation* **19** 577-593.
[19] Daubechies I, Defrise M and De Mol C 2004 *Comm. on Pure and Applied Math.* **57** 1413-1457.
[20] Schuster T, Kaltenbacher B, Hofmann B, and Kazimierski K S, 2010 *Regularization methods in Banach spaces*. Radon Series on Computational and Applied Mathematics.
[21] Hein T and Kazimierski K S 2010 *Inv. Probl.* **26** 055002.
[22] Livesay D E and Chen K M 1974 *IEEE Trans. Microwave Theory Tech.* **MTT-22** 1273-1280.
[23] Schöpfer F, Louis A K, and Schuster T, 2006 *Inv. Probl.* **22** 311-329.
[24] Xu X, Liu Q H and Zhang Z Q 2002 *Proc. IEEE Int. Symp. Antennas and Prop. Soc.* **2** 614-617.

An iterative algorithm for L1-TV constrained regularization in image restoration.

K. Chen[1], **E. Loli Piccolomini**[2], **F. Zama**[2]

[1]Department of Mathematics, University of Liverpool
[2]Department of Mathematics, University of Bologna

E-mail: K.Chen@liverpool.ac.uk, elena.loli@unibo.it, fabiana.zama@unibo.it

Abstract. We consider the problem of restoring blurred images affected by impulsive noise. The adopted method restores the images by solving a sequence of constrained minimization problems where the data fidelity function is the ℓ_1 norm of the residual and the constraint, chosen as the image Total Variation, is automatically adapted to improve the quality of the restored images. Although this approach is general, we report here the case of vectorial images where the blurring model involves contributions from the different image channels (cross channel blur). A computationally convenient extension of the Total Variation function to vectorial images is used and the results reported show that this approach is efficient for recovering nearly optimal images.

1. Introduction

We consider here the problem of restoring a multichannel image Y^δ corrupted by blur and *salt and pepper* noise. The problem is solved by a variational approach, minimizing an objective functional where the data fidelity term is weighted by a suitable regularization term. This formulation requires the choice of the regularization parameter which is very critical. Therefore we consider the constrained formulation where the minimization of the data fidelity term is constrained by a suitably defined regularization term. *Salt and pepper* noise is considered an outlier which do not obey to the Gaussian noise model. Minimization of outliers effects can be obtained by replacing the ℓ_2 in the data fidelity term by the ℓ_1 norm [1, 2, 3, 4]. Moreover the combination of ℓ_1 data fidelity term with edge preserving regularizers has proven to be effective [2, 5, 6, 7, 8]. In this paper, we adopt the Total Variation regularizer extended to Multichannel images (MTV) [9]. We solve the deblurring-denoising problem by the following sequence of constrained minimization problems:

$$min_U \|\mathcal{H}U - Y^\delta\|_1 \;\; s.t. \;\; MTV(U) \le \gamma_j, \quad j = 0, 1, \ldots \tag{1}$$

where $\mathcal{H}$ is the blurring kernel (here $\mathcal{H} \neq I$), U is the exact image (unknown) to be recovered. The initial smoothing term γ_0 is obtained by the input data Y^δ by a filtering procedure specific for multichannel images. The successive terms of the sequence $\{\gamma_j\}$, $j > 0$, are updated by an iterative method based on the residuals of (1). In this sense, the proposed method does not depend on any regularization or constraint parameter and only the recorded data Y^δ are necessary as input. The problems (1) are solved in Lagrangian dual form by the algorithm `CL1TV` presented in section 2. Some numerical experiments are reported in section 3, where the

Journal of Physics: Conference Series **657** (2015) 012009 doi:10.1088/1742-6596/657/1/012009

tests are performed on deblurring of color images, corrupted by cross-channel blur and salt and pepper noise. Finally the conclusions are given in section 4.

2. The Constrained L1-TV Algorithm (CL1TV)

We now discuss our iterative algorithm, named the Constrained L1-TV Algorithm (CL1TV), that computes a suitable set of values γ_j and solves the constrained problem (1) using the Lagrangian dual form. We shall present the algorithm for the case of multichannel images with cross channel blur.

2.1. Multichannel image restoration

Assume our observed p-channel image (with blur) is $Y = [Y_1, Y_2, \ldots, Y_p]$, which is the sum of convolutions of the blurring kernels $H_{n,k}$ with the image channels U_k of the original image $U = [U_1, \ldots, U_p]$:

$$Y_n = \sum_{k=1}^{p} H_{n,k} * U_k, \quad n = 1, \ldots, p \tag{2}$$

where each channel $Y_n \in \mathcal{R}^{M\times N}$, $n = 1, \ldots, p$. For example, the blurring matrix of truecolor images $\mathcal{H}$ has 3×3 blocks ($p = 3$) where the diagonal blocks $H_{k,k}$, $k = 1 \ldots, 3$ represent within-channel blur, while the blocks $H_{n,k}$ $n \neq k$ represent the cross-channel blur, i.e. how the blur on the n-th channel influences the k-channel. For a general p-channel image, we consider discrete Multichannel Total Variation, as extension of the Total Variation for a grayscale image, defined as follows:

$$MTV(U) = \sum_{j=1}^{M}\sum_{i=1}^{N} \left(|\nabla U|^2_{i,j,1} + |\nabla U|^2_{i,j,2} + \cdots + |\nabla U|^2_{i,j,p}\right)^{1/2}.$$

Forward difference formulas are used in $|\nabla U|^2 = U_x^2 + U_y^2$ to approximate the derivatives.

2.2. Computation of the sequence γ_j

The starting value γ_0 of the sequence of regularization constraints $\gamma_j, j = 1, \ldots,$ is defined as the Multichannel Total Variation of a low pass filtered version $U^{(F)}$ of the blurred noisy data. Such a filtered image is computed by solving the following unconstrained problem:

$$U^{(F)} = argmin_U \|\mathcal{H}U - Y^\delta\|_1 + \lambda_F MTV(U), \tag{3}$$

where λ_F is obtained by undersampling the sum of the spectra of the blurring matrices $H_{n,k}$ acting on each channel and taking the minimum of the Fourier coefficients. In the case of cross channel blur, we assume that each submatrix $H_{i,j}$ can be diagonalized by a unitary Fourier matrix F: $H_{i,j} = F^*\Psi^{(i,j)}F$, $i, j = 1, \ldots p$ and we approximate $\mathcal{H}$ by a block diagonal matrix $\bar{\mathcal{H}}$ where each diagonal block is

$$\bar{H}_{i,i} = F^*\bar{\Psi}^{(i)}F, \quad \bar{\Psi}^{(i)} = \frac{1}{p}\sum_{j=1}^{p}\Psi^{(i,j)}, \quad i = 1, \ldots, p.$$

By undersampling the power spectrum we compute the filter parameters β_i as:

$$\beta_i = \min_{1\leq k\leq N} \left|\frac{(FY_i^\delta)_{k\cdot M}}{\bar{\Psi}^{(i)}_{k\cdot M}}\right|, \quad i = 1, \ldots, p \tag{4}$$

and define $\lambda_F = \min_{1\leq i\leq p}\beta_i$. For each channel i, we have that $|(FY_i^\delta)_\ell|$ decay faster than $|(\bar{\Psi}^{(i)})_\ell|$ ($\ell = 1, \ldots, k \cdot M$) until they level off when the noise starts to dominate [10]. Hence

Journal of Physics: Conference Series **657** (2015) 012009 doi:10.1088/1742-6596/657/1/012009

we take the minimum value in (4) to limit as much as possible the influence of noise. Then $\gamma_0 = MTV(U^{(F)})$.

We experimentally observe that the residual function $r(\gamma) = \|\mathcal{H}U(\gamma) - Y^\delta\|_1$ decreases for increasing values of γ (for γ less than an optimal value $\tilde{\gamma}$). In particular $r(\gamma)$ steeply decreases when $\gamma \ll \tilde{\gamma}$ and becomes flat around $\tilde{\gamma}$. We can define an increasing sequence $\gamma_j, j = 1, \dots r$, and compute γ_j by applying a kind of cross validation method, minimizing the residual $r(\gamma)$. In practice, checking the slope of the residual curve, we compute γ_j with the following relation:

$$\gamma_j = \gamma_{j-1}(1+P), \quad \texttt{if } |r(\gamma_j) - r(\gamma_{j-1})| > P_{tol}|(\gamma_j - \gamma_{j-1})| \tag{5}$$

where P_{tol} is an assigned tolerance. Let γ_s be the exit value of (5) then, if the slope is still negative, the adaptation procedure is stopped ($\gamma_r = \gamma_s$). Otherwise, if the slope in γ_s is greater than P_{tol}, then a backtracking procedure (BCKTRK) is performed in the interval $[\gamma_{s-1}, \gamma_s)$, by using (5) with $P = P/2$ and $j = s$.

Algorithm 2.1 (OUTPUT: γ_r).

Compute λ_F *and compute* $U^{(F)}$ *by solving* (3). $\gamma_0 = MTV(U^{(F)})$. *j=0;*

```
repeat
      j = j + 1;   γ_j = γ_{j-1}(1 + P)
      D_j = (r(γ_j) − r(γ_{j-1}))/(γ_j − γ_{j-1})
until D_j < −P_tol
if D_j < P_tol
      γ_r = γ_j
else
      a = γ_{j-1}, b = γ_j
      compute γ_r = BCKTRK(a, b).
end
```

In our experiments, $P_{tol} = 10^{-4}$, $P = 0.1$.

2.3. Solution of the L1-TV subproblem

For each $\gamma = \gamma_j$ the constrained problem (1) is solved using its Lagrangian dual form:

$$\max_{\lambda} \min_{U} \mathcal{L}(U, \lambda), \quad \mathcal{L}(U, \lambda) \equiv \|\mathcal{H}U - Y^\delta\|_1 + \lambda \left(MTV(U) - \gamma\right). \tag{6}$$

Imposing the first order conditions $\nabla_\lambda \mathcal{L}(U, \lambda) = 0$, we compute the solution $(\hat{\lambda}, \hat{U})$ of (6) by solving:

$$\text{find } \hat{\lambda} \text{ s.t. } MTV(\hat{U}(\lambda)) - \gamma = 0 \quad \text{where } \hat{U} \equiv U(\hat{\lambda}) \text{ is the solution of}$$
$$\min_U \|\mathcal{H}U - Y^\delta\|_1 + \lambda \left(MTV(U) - \gamma\right) \tag{7}$$

The monotonicity property of $MTV(U(\lambda)) - \gamma$ allows us to solve the nonlinear equation:

$$MTV(U(\lambda)) - \gamma = 0 \tag{8}$$

by a hybrid bisection+secant method [11] yielding a sequence $\{\lambda_k\}$ converging to the root $\hat{\lambda}$. Usually few bisection iterations are necessary ($k_s \simeq 3$) to guarantee the convergence of the secant iterations and globally less than 10 iterations are performed by the hybrid method when stopped by the following criterion:

$$|MTV(U^{(k)}) - \gamma| < \tau_r |MTV(U^{(0)}) - \gamma| + \tau_a \quad \text{or} \quad |\lambda_k - \lambda_{k-1}| < \tau_a \quad \text{or} \quad k > \texttt{maxit} \tag{9}$$

with $\tau_r = \tau_a = 10^{-3}$ and `maxit` = 15 in our experiments. By solving (7) with $\lambda = \lambda_k$, we obtain a sequence $U^{(k)} = U(\lambda_k)$ converging the solution $\hat{U}$. Using MTV an efficient and fast method for solving (7) is the fast alternating minimization algorithm proposed in [9] (FTVD4). However, different solvers could be applied to (7), related to different extensions of the TV to multichannel images. The procedure for solving a subproblem (1) is outlined in Algorithm 2.2.

Algorithm 2.2 (CL1TV – Input: γ_j, $\mathcal{H}, Y^\delta$, λ_0,k_s; output: U, λ).

compute $U^{(0)}$ *solving* (7) *with* $\lambda = \lambda_0$ *using FTVD4 function*
$k = 0$ *% Solution Computation*
`repeat`
  $k = k + 1$
  Apply a hybrid bisection-secant method for solving $MTV(U(\lambda_k)) - \gamma = 0$
  compute $U^{(k)}$ *solving* (7) *with* $\lambda = \lambda_k$ *using FTVD4 function*
`until` *exit condition* (9)
$U = U^{(k)}$.

3. Numerical Results

In this section we present some numerical results obtained with the proposed `CL1TV` method in the deblurring of color images. The tests have been performed in Matlab 2012a. In all the experiments, the maximum number of allowed `CL1TV` iterations is 5 and the maximum number of bisection-secant iterations is 10. The blurred noisy images Y^δ are obtained by applying the blurring kernel $\mathcal{H}$ to the test image U^* and by adding salt and pepper noise. In our experiments, we consider two levels of noise: medium noise and high noise, where the number of corrupted pixels in Y^δ is 40% and 80%, respectively. The following cross-channel blurring matrix has been used:

$$\mathcal{H} = \begin{pmatrix} .7G(21,11) & .15G(21,11) & .15G(21,11) \\ .1G(21,11) & .8G(21,11) & .1G(21,11) \\ .2G(21,11) & .2G(21,11) & .6G(21,11) \end{pmatrix}. \tag{10}$$

where $G(h,\sigma)$ is a Gaussian function of size $h = 21$ pixels and variance $\sigma = 11$ pixels. The quality of the reconstructed images is evaluated through the relative error: $E = \|U^* - \tilde{U}\|_2 / \|U^*\|_2$ where U^* and $\tilde{U}$ are the exact and reconstructed image, respectively. We show the results on two color images:

- the I1 photographic image in figure 1;
- the I2 synthetic image of figure 2.

(a) Original (b) Noisy blurred (high noise) (c) Restored

Figure 1: Original, blurred (high noise) and restored images I1.

In figure 3 we report the relative errors (figures (a) and (c)) and the L1 residuals (figures (b) and (d)) as functions of the parameter γ for I1. We have obtained the plots by considering a

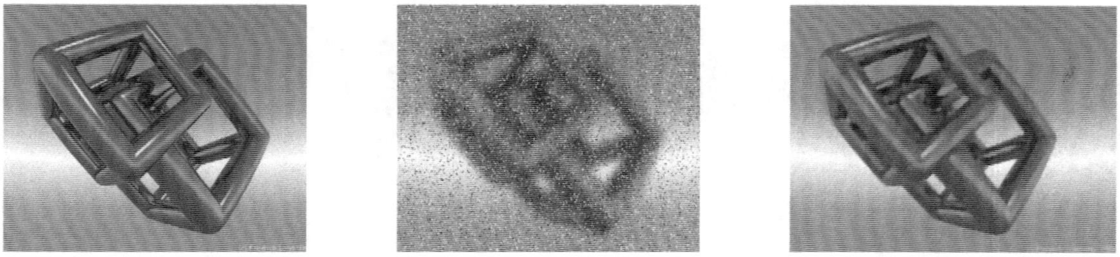

(a) Original (b) Noisy blurred (medium noise) (c) Restored

Figure 2: Original, blurred (medium noise) and restored images I2.

set of equispaced values γ_i in the interval $[8.10^3, 2.10^4]$ (sufficiently larger than $[\gamma_0, \gamma_s]$) and by solving problems of the form (1) with the proposed `CL1TV` method. From the plots (a) and (c) (medium and high noise, respectively) we see that the computed solution is very close to the best possible with the considered model. This means that the estimate of γ made by our algorithm is good. Figures (b) and (d) show that the residual really decreases and then becomes flat near the best value of γ; confirming the validity of method (5) based on the computed residual $r(\gamma)$. We remind that the relative errors reported in figures (a) and (c) are not used in the algorithm for choosing γ, but they are shown only to confirm the validity of the proposed method. In table 1 we report some numerical results obtained with the `CL1TV` method. It is evident that the relative error is strongly reduced from the recorded image ($E0$) to the computed image (Er). The computational time is proportional to the number of FTVD iterations (implementing the Alternating Directions algorithms).

Test (nl)	$E0$	Er	it_γ	it_λ	it_{FTVD}
I1 40%	0.477	0.139	5	33	15768
I1 80%	0.600	0.145	5	24	10941
I2 40%	0.306	0.050	5	37	5670
I2 80%	0.4166	0.051	5	27	4021

Table 1: `CL1TV` algorithm results. $E0$ is the relative error of the recorded image, Er the relative error of the restored image, it_γ the number of elements of the sequence γ_i, it_λ the total number of iterations of the bisection and secant methods, it_{FTVD} the total number of iterations of the FTVD method.

4. Conclusions

In this paper we proposed the `CL1TV` iterative method for automatic restoration of blurred noisy images corrupted by impulsive noise. The method only requires the recorded image and the blurring kernel as input. The algorithm solves a sequence of constrained minimization problems where the objective function is the ℓ_1-norm of the residual and the constraint function is the Total Variation (extended for vectorial images). The experiments carried on some images proved that the method efficiently recovers nearly optimal results; hence they are encouraging for further tests in deblurring and other linear inverse problems in imaging, such as tomographic

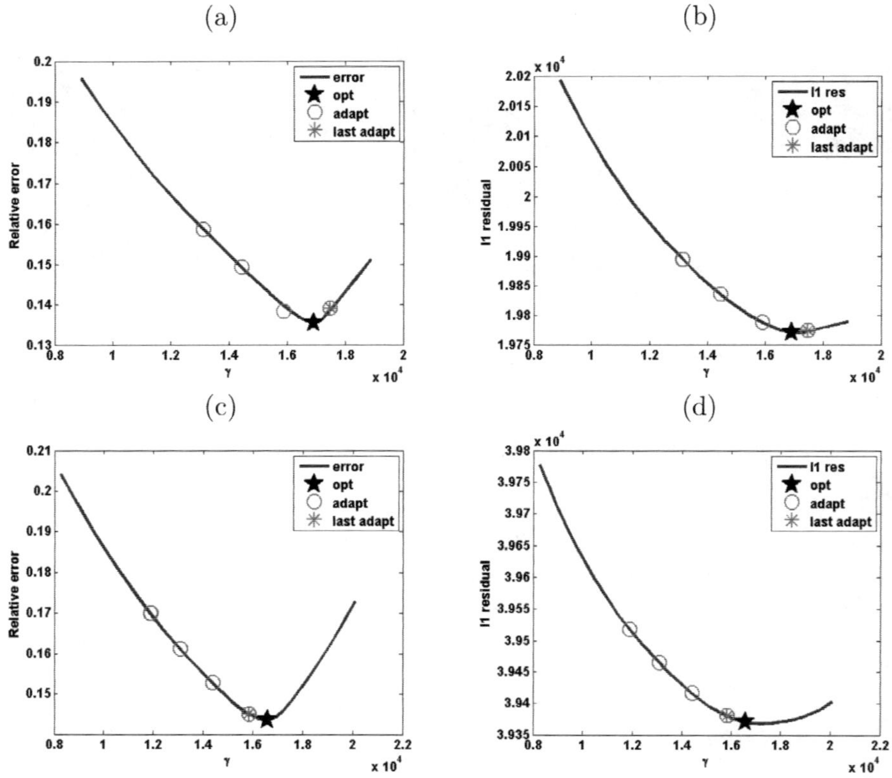

Figure 3: Relative errors (a)(c) and L1 residuals (b)(d) for the tests with medium 40% noise (a)(b) and high 80% noise (c)(d).

reconstruction. In future work we will compare our approach to other methods proposed in the literature and test different algorithms for the solution of the unconstrained problem (7).

Acknowledgement
The authors would like to thank the reviewers for their comments and suggestions.

References

[1] Nikolova M 2004 *Journal of Mathematical Imaging and Vision* **20** 99–120
[2] Chan R, Ho C W and Nikolova M 2005 *Image Processing, IEEE Transactions on* **14** 1479–1485
[3] Bar L, Brook A, Sochen N and Kiryati N 2007 *Image Processing, IEEE Transactions on* **16** 1101–1111
[4] Chan R H and Chen K 2008 *SIAM Journal on Scientific Computing* **30** 1474–1489
[5] Chan T and Esedoglu S 2006 *Siam J. Appl. Mathem.* **66** 1632–1648
[6] Karkkainen T, Kunish K and Majava K 2005 *Computing* **74** 353–376
[7] Clason C, Jin B and Kunish K 2010 *Siam J. Sci. Comput.* **32** 1484–1505
[8] Condat L and Mosaddegh S 2012 Joint demosaicking and denoising by total variation minimization *Image Processing (ICIP), 2012 19th IEEE International Conference on* pp 2781–2784
[9] Yang J, Zhang Y and Yin W 2009 *Siam J. Sci. Comput.* **31** 2482–2865
[10] Hansen P C, Nagy J and O'Leary D 2006 *Deblurring Images: Matrices, Spectra, And Filtering* (SIAM)
[11] Chen K, Piccolomini E L and Zama F 2014 *Numerical Algorithms* **67** 73–92

Journal of Physics: Conference Series **657** (2015) 012010 doi:10.1088/1742-6596/657/1/012010

A comparative assessment of information-exploitation techniques for GPR data inversion

M Salucci[1], L Tenuti[2], L Poli[2], G Oliveri[2] and A Massa[1,2]
[1] ELEDIA Offshore Lab @ Paris, Laboratoire des Signaux et Systèmes (L2S, UMR CNRS 8506), 3 rue Joliot Curie, 91192 Gif-sur-Yvette, France
[2] ELEDIA Research Center, Department of Information Engineering and Computer Science, University of Trento, Via Sommarive 5, 38123 Trento, Italy
E-mail: andrea.massa@l2s.centralesupelec.fr

Abstract. The inversion of Ground Penetrating Radar (GPR) data requires the development of suitable information-exploitation techniques that are able to extract as much as possible information on the unknown targets from the available measurements. An innovative single-frequency (SF) inversion technique based on a deterministic conjugate-gradient (CG) minimization and the iterative multi-scaling approach (IMSA) is described. It is then shown how to improve the performances of the SF-IMSA-CG method by the introduction of an external frequency hopping (FH) iterative loop. On the one hand, the proposed FH-IMSA-CG method allows to exploit the intrinsic frequency diversity of wideband GPR measurements thanks to the FH strategy. On the other hand, the IMSA approach guarantees a significant reduction of the problem unknowns, providing an increased resolution within the identified regions of interest (RoIs). A numerical comparison shows the advantages of the FH-IMSA-CG over its single-frequency version. Moreover, the benefits of integrating the IMSA within the FH are verified by directly comparing the FH-IMSA-CG with its single-resolution (BARE) version (FH-BARE-CG).

1. Introduction

Ground penetrating radar (GPR) is one of the most widespread instrumentations for performing subsurface investigations in many fields of applications, ranging from civil engineering to archaeology [1]. Recently, an increasing amount of research has been focused on the development of fast, effective and robust algorithms for suitably processing the scattered radiation (i.e., the GPR radargram) and estimating the geometrical and dielectric characteristics of unknown objects embedded inside the inspected (inaccessible) domain. As a matter of fact, the reduced amount of collectable information due to the availability of aspect-limited measurements (i.e., often the antennas are placed only above the interface) and to the occurrence of non-negligible reflections in crossing heterogeneous media makes the retrieval of buried targets a quite challenging task. Following the above considerations, suitable information-exploitation techniques are needed to properly address the inversion of GPR data, both considering deterministic [2] and stochastic [3] approaches. When some *a-priori* information on the scenario at hand is available, such a task may be accomplished by introducing some mathematical approximations that allow to reduce the non-linearity of the problem (e.g., in the presence of weak scatterers [2, 4]). Similarly, compressive sensing (CS)-based techniques [4-6] have been proven to be very effective if the retrieved solution is sparse with respect to a properly chosen representation basis. Iterative multi-scaling approaches (such as, for example, the IMSA method [7]) can be successfully employed to effectively tackle both the *non-linearity* and the *ill-posedness* of the subsurface inversion problem [2], leading to an increased resolution within the identified regions of interest (RoIs). Moreover, the intrinsic wideband nature of GPR data enables the use of techniques that are able to process different components of the measured spectrum, allowing to exploit the frequency-diversity as an additional source of information [5].

In this framework, this work presents an innovative imaging technique based on the integration of a deterministic conjugate-gradient (CG) method and the IMSA that is able to retrieve buried targets by processing a single-frequency (SF) component of the GPR spectrum. It is then shown how to enclose the SF-IMSA-CG method within an external frequency-hopping (FH) loop in order to exploit the additional information coming from multi-chromatic components of the data. A numerical assessment is given in order to verify the advantages of the FH-IMSA-CG method over the SF-IMSA-CG. Moreover, the benefits of integrating the IMSA within the FH are verified by directly comparing the FH-IMSA-CG with its single-resolution (BARE) version (FH-BARE-CG).

2. Mathematical formulation

Let be given a 2-D scenario where the investigation domain D_{inv} is completely embedded inside a lossy homogeneous background with relative permittivity ε_{rb} and conductivity σ_b (Fig. 1).

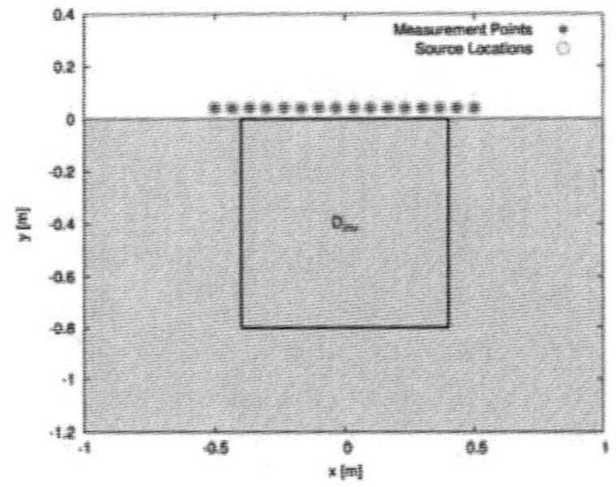

Figure 1. Geometry of the problem.

A set of V z-directed ideal line sources are located in the upper half-space (free-space) at constant height h above the interface ($y = 0$) and are radiating a wide-band transverse-magnetic field $\mathbf{\Psi}^v_{inc}(\mathbf{r},t) = \Psi^v_{inc}(\mathbf{r},t)\hat{\mathbf{z}}$, $\mathbf{r} = (x, y)$ being the position vector. Moreover, let us assume that an unknown cylindrical scatterer with relative permittivity $\varepsilon_r(\mathbf{r})$ and conductivity $\sigma(\mathbf{r})$ is buried at a given depth inside D_{inv}. Under these hypothesis, the interactions between the v-th illumination and the lower half-space produce a total field $\mathbf{\Psi}^v_{tot}(\mathbf{r},t) = \{\Psi^v_{inc}(\mathbf{r},t) + \Psi^v_{scatt}(\mathbf{r},t)\}\hat{\mathbf{z}}$ that is collected by M ideal field probes located above the interface at positions $\mathbf{r}^v_m$ ($m = 1,\ldots,M$, Fig. 1), $\Psi^v_{scatt}(\mathbf{r},t)$ being the time-domain field scattered by the buried target. By means of a Fast Fourier Transform (FFT) algorithm, the received radiation is transformed into the frequency domain, in which the scattering phenomena are completely modelled at a given frequency by means of the following set of non-linear equations

$$\begin{aligned} E^v_{scatt}(\mathbf{r}^v_m) &= k_b^2 \int_{D_{inv}} \tau(\mathbf{r}') E^v_{tot}(\mathbf{r}') G_{ext}(\mathbf{r}^v_m, \mathbf{r}') d\mathbf{r}' \\ E^v_{inc}(\mathbf{r}) &= E^v_{tot}(\mathbf{r}) - k_b^2 \int_{D_{inv}} \tau(\mathbf{r}') E^v_{tot}(\mathbf{r}') G_{\mathrm{int}}(\mathbf{r}, \mathbf{r}') d\mathbf{r}' \end{aligned} \tag{1}$$

where k_b is the wave-number in the lossy medium, G_{ext} and G_{int} are the external and internal Green's functions for the half-space, while $\tau(\mathbf{r})$ is the so-called contrast function and is defined as

$$\tau(\mathbf{r}) = \{\varepsilon_r(\mathbf{r}) - \varepsilon_{rb}\} + j\left\{\frac{\sigma_b - \sigma(\mathbf{r})}{2\pi f \varepsilon_0}\right\}. \tag{2}$$

The SF-IMSA-CG method extracts a single component of the transformed GPR spectrum and recovers an estimation of the unknown contrast and total field inside D_{inv} in S iterative steps. This method can be summarized as follows.

1. *Initialization*. At the first multi-scaling step ($s = 1$) D_{inv} is partitioned into N square sub-domains (N being the number of degrees-of-freedom of the scattered field [8]). The unknowns are initialized to the empty background [i.e., $\tau^{(0)}(\mathbf{r}_n) = 0.0$ and $E_{tot}^{v(0)}(\mathbf{r}_n) = E_{inc}^{v}(\mathbf{r}_n)$, for $n = 1,...,N$ and $v = 1,...,V$].
2. *Preliminary coarse reconstruction*. A CG deterministic approach is used to recover an initial low-resolution reconstruction by minimizing the following cost function

$$\Phi = \frac{\sum_{v=1}^{V}\sum_{m=1}^{M}\left|E_{scatt}^{v}(\mathbf{r}_m^v) - \widetilde{E}_{scatt}^{v}(\mathbf{r}_m^v)\right|^2}{\sum_{v=1}^{V}\sum_{m=1}^{M}\left|E_{scatt}^{v}(\mathbf{r}_m^v)\right|^2} + \frac{\sum_{v=1}^{V}\sum_{n=1}^{N}\left|E_{inc}^{v}(\mathbf{r}_n) - \widetilde{E}_{inc}^{v}(\mathbf{r}_n)\right|^2}{\sum_{v=1}^{V}\sum_{n=1}^{N}\left|E_{inc}^{v}(\mathbf{r}_n)\right|^2} \tag{3}$$

   where $\widetilde{E}_{scatt}^{v}(\mathbf{r}_m^v)$ and $\widetilde{E}_{inc}^{v}(\mathbf{r}_n)$ denote the computed scattered and incident field, respectively.
3. *Iterative multi-scaling loop*. At each s-th successive step ($s = 2,...,S$), a clustering procedure [7] is applied to identify the RoI $D^{(s)}$ where the target has been localized at the previous step. An higher resolution image of $D^{(s)}$ is then obtained by means of the CG, by suitably mapping the solution found at the $(s-1)$-th step to properly initialize the problem unknowns.
4. *Termination*. The multi-scaling procedure is terminated when $s = S$ or when a suitable stopping criterion is met [7].

Given the intrinsic wide-band nature of GPR data, the performances of the SF-IMSA-CG can be improved by nesting it inside a FH external loop. The resulting FH-IMSA-CG method extracts K frequency components of the measured spectrum inside the 3dB bandwidth of the radiated pulse (i.e., $f_{\min} \le f_1 \le ... \le f_K \le f_{\max}$). Then, each k-th frequency component is successively processed by means of the described algorithm in a cascaded fashion. At each frequency stage successive to the first one (i.e., $k = 2,...,K$), the solution found at the IMSA convergence step S_{k-1} of the $(k-1)$-th stage is used to initialize the solution as follows

$$\begin{aligned} \tau_k^{(0)}(\mathbf{r}) &= \mathrm{Re}\left\{\tau_{k-1}^{(S_{k-1})}(\mathbf{r})\right\} + j\frac{f_{k-1}}{f_k}\mathrm{Im}\left\{\tau_{k-1}^{(S_{k-1})}(\mathbf{r})\right\} \\ E_{tot,k}^{(0)v}(\mathbf{r}) &= F\left\{\tau_k^{(0)}(\mathbf{r}); E_{inc,k}^{v}(\mathbf{r})\right\} \end{aligned} \tag{4}$$

where $F\{.\}$ indicates a 2-D method-of-moments (MoM) forward solver.

3. Numerical comparison

This section is aimed at assessing the potentialities, as well as the limits, of the proposed techniques when dealing with the reconstruction of unknown buried targets. In order to give a quantitative measure of the accuracy of the retrieved profiles, the same reconstruction errors defined in [9] are used hereinafter. Moreover, in order to test the algorithms under realistic noisy conditions, the measured total field has been corrupted in time-domain by an additive zero-mean Gaussian noise $n^v(\mathbf{r}_m^v;t)$, by considering the following definition of signal-to-noise ratio

$$SNR = 10\log_{10} \sum_{v=1}^{V}\sum_{m=1}^{M}\int_{-\infty}^{+\infty}\left|\Psi_{tot}^{v}\left(\mathbf{r}_m^v;t\right)\right|^2 dt \Big/ \sum_{v=1}^{V}\sum_{m=1}^{M}\int_{-\infty}^{+\infty}\left|n^{v}\left(\mathbf{r}_m^v;t\right)\right|^2 dt \tag{5}$$

3.1. Homogeneous scatterer at different depths

In order to assess the effectiveness of the proposed inversion techniques, let us consider a square investigation domain of side 0.8 m (Fig. 1), with $\varepsilon_{rb} = 4.0$ and $\sigma_b = 10^{-3}$ S/m.

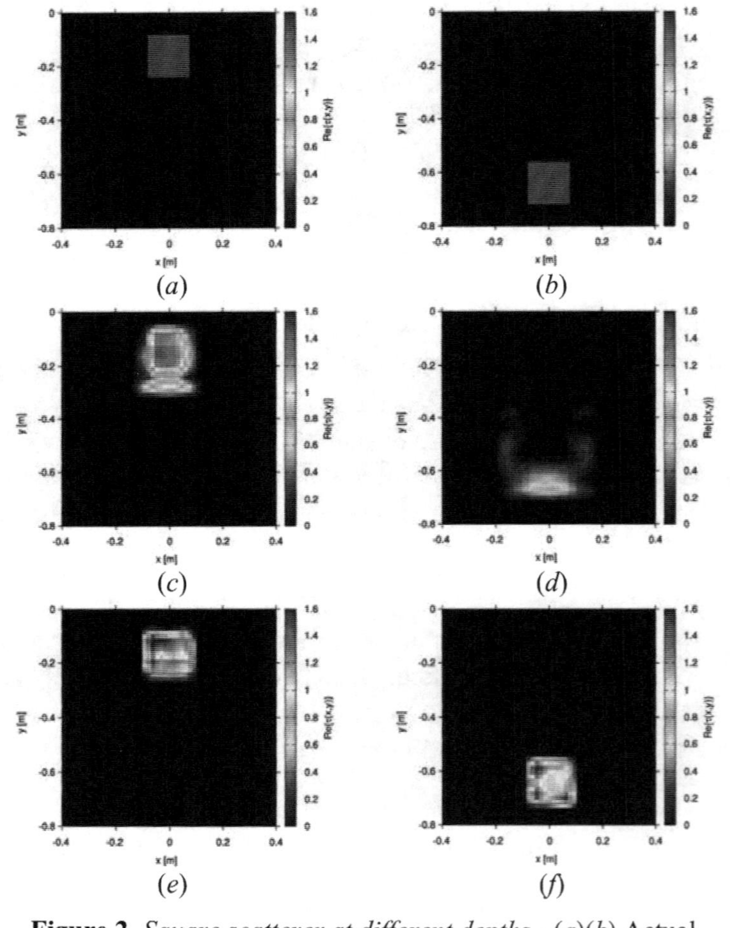

Figure 2. *Square scatterer at different depths* - (*a*)(*b*) Actual dielectric profiles and reconstructions by (*c*)(*d*) SF-IMSA-CG and (*e*)(*f*) FH-IMSA-CG, for *SNR* = 40 dB.

The synthetic field data have been generated by means of the GprMax2D simulator [10], by considering $V = 16$ sources equally spaced along a line of length 1 m located at $h = 0.04$ m above the interface. A Gaussian monocycle pulse with a central frequency of $f_0 = 300$ MHz and 3dB bandwidth covering the [200, 600] MHz range is used as excitation signal. For each illumination the scattered radiation is collected by a set of $M = 15$ probes co-located with the sources. This first analysis deals with a benchmark square-shaped homogeneous scatterer ($\tau = 1.5$) of side 0.16 m buried at different depths inside D_{inv}. More precisely, we suppose that the target barycentre is located at $(x_c, y_c) = (0.0, -0.16)$ m [Fig. 2(*a*)] and $(x_c, y_c) = (0.0, -0.64)$ m [Fig. 2(*b*)], respectively. The retrieved profiles by the SF-IMSA-CG are shown in Figs. 2(*c*)-2(*d*), when considering a single frequency component corresponding to f_0 and $SNR = 40$ dB. As it can be easily noticed, the quality of the reconstructions significantly degrades as the depth of the target is increased [Fig. 2(*c*) vs. Fig. 2(*d*)]. Such a consideration is confirmed by the computed internal error [i.e., $\Xi_{\mathrm{int}}^{SF-IMSA-CG} = 1.74 \times 10^{-1}$ for Fig. 2(*c*) and $\Xi_{\mathrm{int}}^{SF-IMSA-CG} = 4.21 \times 10^{-1}$ for Fig. 2(*d*)].

On the other hand, Figs. 2(*e*)-2(*f*) show the reconstructions obtained by the FH-IMSA-CG approach, considering $K = 5$ equally-spaced frequency components within the [200, 600] MHz bandwidth. As expected, a non-negligible improvement in terms of reconstruction accuracy is observed with respect to the SF approach, thanks to the exploitation of the additional information coming from multi-frequency GPR data. The advantage of using the FH strategy becomes relevant when dealing with the retrieval of the deepest object [Fig. 2(*f*) vs. Fig. 2(*d*)]. In this case, the FH-IMSA-CG is able to provide a better reconstruction of the object shape, as well as of its dielectric characteristics, as denoted by a significantly lower internal error ($\Xi_{\mathrm{int}}^{FH-IMSA-CG} = 2.21 \times 10^{-1}$).

3.2. *Inhomogeneous scatterer*

The next numerical benchmark considers an inhomogeneous scatterer [Fig. 3(*a*)] characterized by an inner core with $\tau_{\mathrm{int}} = 2.0$ surrounded by an external region with $\tau_{ext} = 1.0$.

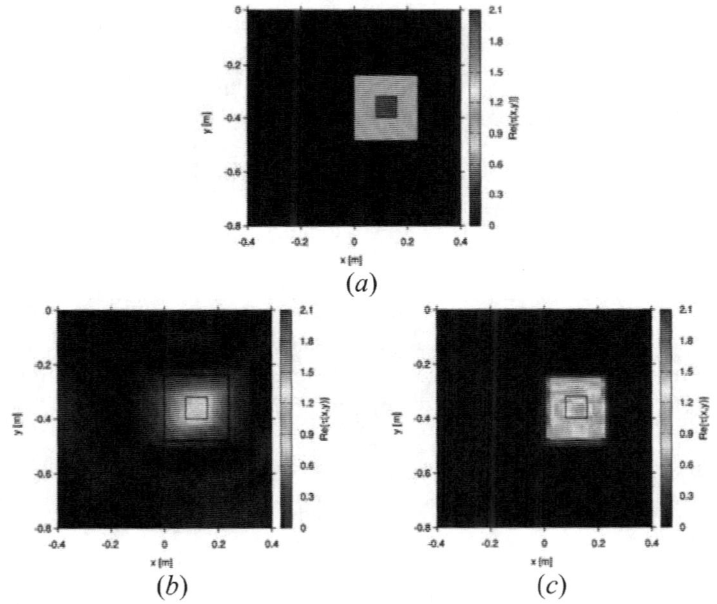

Figure 3. *Inhomogeneous scatterer* - (*a*) Actual dielectric profile and retrieved profiles by (*b*) FH-BARE-CG and (*c*) FH-IMSA-CG for *SNR* = 50 dB.

All the parameters are the same of the previous analysis, except from the number of sources and measurement points, which has been set to $V = 20$ and $M = 19$, respectively. The retrieved profiles for both the single-resolution FH-BARE-CG [Fig. 3(*b*)] and the multi-resolution FH-IMSA-CG [Fig. 3(*c*)] are shown in Fig. 3, when $SNR = 50$ dB. As it can be observed, the FH-IMSA-CG is not only able to suppress the undesired artefacts in the background region, but it also improves the overall quality of the reconstruction, by correctly identifying the presence of an higher contrast within the inner core of the target, as further confirmed by the reconstruction error [$\Xi_{tot}^{FH-IMSA-CG} = 1.62\times10^{-2}$ for Fig. 3(*c*) vs. $\Xi_{\text{int}}^{FH-BARE-CG} = 7.53\times10^{-2}$ for Fig. 3(*b*)]. Thanks to a reduced number of unknowns, a remarkable improvement is observable also when considering the computational costs (i.e., $\Delta t^{FH-IMSA-CG} \approx 2.5\times10^{4}$ s vs. $\Delta t^{FH-BARE-CG} \approx 7.0\times10^{4}$ s on a standard laptop with 4 GB of RAM memory).

4. Conclusions

A comparative assessment between different information-exploitation techniques for the inversion of GPR data has been presented. It has been shown that the proposed FH-IMSA-CG method is able to overcome, in terms of reconstruction accuracy, the SF-IMSA-CG method thanks to its ability to exploit the additional information coming from multiple frequency components of the measured GPR spectrum. The advantages of the FH-IMSA-CG over its single-resolution version (FH-BARE-CG) have been also proved by means of a representative numerical example. The use of the IMSA approach has allowed a significant reduction of the problem unknowns, providing an increased resolution within the identified RoIs. Moreover, a remarkable improvement in terms of computational times has been obtained.

References

[1] Leckebusch J 2003 Ground-penetrating radar: a modern three-dimensional prospection method *Archaeological Prospection* **10** 213-240

[2] Salucci M, Sartori D, Anselmi N, Randazzo A, Oliveri G and Massa A 2013 Imaging buried objects within the second-order Born approximation through a multiresolution-regularized inexact-Newton method *Int. Symp. Electromag. Theory (Hiroshima, Japan)* 116-118

[3] Rocca P, Benedetti M, Donelli M, Franceschini D and Massa A 2009 Evolutionary optimization as applied to inverse problems *Inverse Problems* **25** 1-41

[4] Oliveri G, Anselmi N and Massa A 2014 Compressive sensing imaging of non-sparse 2D scatterers by a total-variation approach within the Born approximation *IEEE Trans. Antennas Propag.* **62** 5157-5170

[5] Poli L, Oliveri G, Ding P-P, Moriyama T and Massa A 2014 Multifrequency Bayesian compressive sensing methods for microwave imaging *J. Opt. Soc. Am.* A **31** 2415-2428

[6] Poli L, Oliveri G and Massa A 2013 Imaging sparse metallic cylinders through a local shape function Bayesian compressive sensing approach *J. Opt. Soc. Am.* A **30** 1261-1272

[7] Caorsi S, Donelli M, Franceschini D and Massa A 2003 A new methodology based on an iterative multiscaling for microwave imaging *IEEE Trans. Microwave Theory Tech.* **51** 1162-1173

[8] Bucci O M, Crocco L, Isernia T and Pascazio V 2001 Subsurface inverse scattering problems: quantifying qualifying and achieving the available information *IEEE Trans. Geosci. Remote Sensing* **39** 2527-2538

[9] Moriyama T, Salucci M, Oliveri G, Tenuti L, Rocca P and Massa A 2014 A Multi-scaling deterministic imaging for GPR survey *Proc. 2014 IEEE Antenna Conf. Ant. Meas. Appl. (Antibes Juan-les-Pins)* 1-3

[10] Giannopoulos A 2005 Modelling ground penetrating radar by GprMax *Construct. Build. Mater.* **19** 755 -762

Acknowledgment

This work has been partially supported by the SIRENA project (2014-2017) funded by DIGITEO (France) under the "Call for Chairs 2014" and benefited from the networking activities carried out within the EU funded COST Action TU1208 "Civil Engineering Applications of Ground Penetrating Radar".

Journal of Physics: Conference Series **657** (2015) 012011 doi:10.1088/1742-6596/657/1/012011

Image regularization for Poisson data

A Benfenati[1] and V Ruggiero[2]

[1 2] Polo Scientifico Tecnologico, Blocco B, Via G. Saragat, 1, 44120 Ferrara (FE)

E-mail: [1]`alessandro.benfenati@unife.it`, [2]`valeria.ruggiero@unife.it`

Abstract. Recently, Poisson noise has become of great interest in many imaging applications. When regularization strategies are used in the so-called Bayesian approach, a relevant issue is to find rules for selecting a proper value of the regularization parameter. In this work we compare three different approaches which deal with this topic. The first model aims to find the root of a *discrepancy* equation, while the second one estimates such parameter by adopting a *constrained* approach. These two models do not always provide reliable results in presence of low counts images. The third approach presented is the inexact Bregman procedure, which allows to use an overestimation of the regularization parameter and moreover enables to obtain very promising results in case of low counts images and High Dynamic Range astronomical images.

1. Introduction

In many imaging applications (e.g. astronomy, emission tomography) the detected image is affected by Poisson noise, due to the fluctuation of the incoming photons hitting the recording electronic device. All the pixels of the detected image $g \in \mathbb{R}^m$ can be considered as realizations of independent and identically distributed Poisson random variables G_i, $i = 1, \dots, m$; in a Bayesian approach [12], the unknown object $x \in \mathbb{R}^n$ is also considered as a realization of a multivalued random variable X and the *a priori* information is encoded into a given probability distribution (*prior*); an estimate of the source image can be obtained by maximizing the so–called posterior probability distribution of X or, equivalently, by solving the following minimization problem

$$\min_{x \geq 0} \varphi_\beta(Hx + b; g) \equiv \varphi_0(Hx + b; g) + \beta\varphi_1(x). \tag{1}$$

Here $\varphi_0(Hx + b; g)$ is the fit–to–data functional, which measures the discrepancy between the detected image g and the unknown image; φ_1 is the regularization function, which is assumed to be non–negative and convex, not necessarily differentiable; the problem is constrained since pixels' values are non–negative. $\beta \geq 0$ is a real parameter which measures the trade–off between φ_0 and φ_1 and it plays a role similar to that of the regularization parameter in Tikhonov theory. For an exhaustive treatment of the statistical model, see for example [5]. The imaging matrix $H \in \mathbb{R}^{m \times n}$ satisfies the assumptions $H_{i,j} \geq 0$, $\sum_{j=1}^n H_{i,j} > 0$; the non–negative term b is a constant background emission. Under the above assumptions, $\{x \geq 0 | Hx + b > 0\} \neq \emptyset$. Due to the presence of Poisson noise, a suitable choice for φ_0 is the generalized Kullback–Leibler divergence [5] $D_{\mathcal{KL}}(y; z) = \sum_{i=1}^{m} \left\{ y_i \log\left(\frac{y_i}{z_i}\right) + z_i - y_i \right\}$, with $y \log(y) = 0$ for $y = 0$. This function is non–negative, convex and coercive on its domain [11].

The quality of the restored image depends strongly on the β value used in (1). Unfortunately,

in presence of Poisson noise, such a value is very hard to estimate: in recent years several methods [2, 6] have been developed for this purpose. We mainly focus on the *Discrepancy Model* presented in [6, 18] and we will refer to it as **Model 1**. Several authors [9, 15], pointing out that such a model can require a considerably high computational effort, developed a *Constrained Model*: we will refer to it as **Model 2**. In [18] a comparison between the two models for some regularization functionals had been performed.
In presence of low counts images, we observe that both models do not always provide satisfactory results. A third approach, the Bregman procedure, which allows us to employ an *overestimation* of β, has been proposed for the Gaussian case [14]. Then, in [8], it was extended to the Poisson case; recently, in [3], in order to achieve a reasonable computational cost, an inexact procedure has been proposed. In this paper the inexact Bregman method is compared with Models 1 and 2 and it is shown that this scheme is very promising in terms of accuracy and efficiency for low counts images and for High Dynamic Range (HDR) images, when one is interested in detecting a weak signal close to very bright point sources.
The paper is organized as follows: in section 2 the basic ideas of the three models are recalled, in section 3 some numerical tests regarding the comparison between such models are presented together with an application concerning HDR images. Finally, the main conclusions are presented in section 4.

2. Descriptions of the models

In this section Model 1, Model 2 and the Bregman procedure are recalled; all the technical details can be found in the references listed in the text. We denote by x^* the true source image corresponding to the detected data g and by $x_\beta \geq 0$ the minimizer of (1) for a given β.

Model 1. In [17, 6] it is proved that when the expected value of each Poisson random variable G_i, $i = 1, \ldots, m$ is sufficiently large, the expected value for $D_{\mathcal{KL}}(g; Hx^* + b)$ is $m/2$. By defining the *discrepancy function* as $\mathcal{D}_H(x; g) = (2/m)D_{\mathcal{KL}}(g; Hx + b)$, the Discrepancy Model consists in finding a value for β such that

$$\mathcal{D}_H(x_\beta; g) = \eta \tag{2}$$

with η close to one. For several regularization functions, the minimizer of (1) exists and it is unique, and since $\mathcal{D}_H(x_\beta; g)$ is an increasing function of β, equation (2) admits one and only solution (see [6, 18]).

Model 2. The *Constrained Model* consists in solving the problem

$$\min_{x \geq 0} \varphi_1(x) \text{ subject to } \mathcal{D}_H(x; g) \leq \tau \tag{3}$$

for a given $\tau > 0$. In this case, only one minimization is required. Model 2 was firstly proposed in [15]. This formulation is linked with the estimation of β since one can write (3) as $\min_{x \geq 0} \varphi_1(x) + \lambda\,(\mathcal{D}_H(x; g) - \tau)$, where, following [6, 17], $\tau = \eta$: for a convenient value of λ (i.e. $\lambda = m(2\beta)^{-1}$), (3) and (2) have the same solutions. Indeed, in [18] under suitable assumptions on φ_1, it is proved that the two models provide the same parameter estimation, even in presence of non–negative data g, a very common situation in practical cases such as microscopy and astronomy.

Inexact Bregman Procedure. The main idea of the Bregman procedure [14, 8] consists in substituting φ_1 with its Bregman distance; the proposed inexact scheme [3] uses, in case of non differentiable regularization term, the ε–subgradients in computing such a distance, introducing the *inexact* Bregman distance. If φ is a convex function, its inexact Bregman distance from x to y is $\Delta_\varepsilon^\xi\varphi(x, y) = \varphi(x) - \varphi(y) - \langle \xi, x - y\rangle + \varepsilon$ with $\xi \in \partial_\varepsilon\varphi(y)$, $\varepsilon \geq 0$, where $\partial_\varepsilon\varphi(y)$ is the ε–subdifferential of φ at the point y. For differentiable regularization term, $\varepsilon = 0$ and the subdifferential is the differential. The inexact Bregman scheme is an iterative procedure which at each step determines the approximate solution of the function $\psi_\beta(x; x^k) \equiv \varphi_0(Hx + b; g) + \beta\Delta_{\varepsilon_k}^{\xi_k}\varphi_1(x, x^k)$, where $\xi^k \in \partial_{\varepsilon_k}\varphi_1(x^k)$. The key idea of the procedure is to solve inexactly the minimization problem $\min \psi_\beta(x; x^k)$, stopping the inner iterative method when $\|\eta_{k+1}\|_2 \leq \mu_{k+1}$ and $\varepsilon_{k+1} \leq \nu_{k+1}$, with $\eta_{k+1} \in \partial_{\varepsilon_{k+1}}\psi_\beta(x; x^k)$, μ_k and ν_k such that $\sum_{i=1}^\infty \mu_i < \infty$, $\sum_{i=1}^\infty i\nu_i < \infty$. Under suitable assumptions [3, Proposition 4], the scheme converges to a minimizer of φ_0. For inverse problems, such as image restoration, the Bregman procedure (exact and inexact version) has the typical semiconvergence behaviour of the iterative methods. Thus, this procedure, coupled with an early stopping criterion, can be used as a regularization technique, without requiring an exact estimation of β. Indeed, a wide experimentation on low and high counts image confirms that an overestimation of β provides reliable results, with an enhancement of the contrast in the restored images and at the same time with a reasonable computational cost, in view of the inexact version of the procedure. This approach seems particularly efficient also in case of HDR images [4], when the Model 1 and the Model 2 do not always provide satisfactory results.

3. Numerical Experiments

In the following we describe a set of test-problems. For each problem, the optimal value β_{opt} is determined by trials, solving (1) for values of β uniformly distributed in an interval $[\beta_{\min}, \beta_{\max}]$; we define β_{opt} as the value corresponding to the minimum relative reconstruction error.

- **spacecraft**: the original image is a 256×256 image with sharp details, with values lying in $[0, 255]$ and the background b is 1; the imaging matrix is taken from `www.mathcs.emory.edu/~nagy/RestoreTools`. The ℓ_2 relative error from the detected data and the original image is 0.709139, β_{opt} is equal to $1.63 \cdot 10^{-3}$.
- **cameraman**: a 256×256 image with values in $[0, 3000]$. The imaging matrix derives from a Gaussian PSF with $\sigma = 1.3$; b is 0 and the ℓ_2 relative distance of g from x^* is 0.116422; β_{opt} is equal to $6.716 \cdot 10^{-3}$.
- **micro**: a confocal microscopy phantom of size 128×128 described in [16], with values in $[0, 69]$; the H matrix is the same used in [16]; b is 0. The relative ℓ_2 relative distance between the x^* and g is 0.190869; β_{opt} is $4.77 \cdot 10^{-2}$.

For all the three models, $\varphi_0 \equiv D_{\mathcal{KL}}$ and φ_1 is the Hyper Surface (**HS**) potential: $\varphi_1(x) = \sqrt{\sum_i \|\nabla_i x\| + \delta^2}$, where ∇_i is the discrete gradient at the i-th pixel and $\delta > 0$ (see [17] and references therein).
Model 1 and inexact Bregman method use as inner solver the Scaled Gradient Projection method [7]. Model 2 is implemented by an Alternating Directions Method of Multipliers, as specified in [15]. An adaptive rule for the selection of the parameter of this method is used (see [18]). In Table 1 the numerical results are reported. For cameraman, an high counts image, the three methods are equivalent, while for micro and spacecraft (low counts images) Model 1 and Model 2 do not achieve good results. Figure 1 shows the obtained restored images.

A common problem rising in astronomical imaging is the restoration of HDR images, i.e. images composed by very bright point sources of known position and faint diffuse objects around such sources. A suitable approach is to consider the signal as the sum of two terms [10], namely the point component x_p and the diffuse component x_d: $x = x_p + x_d$. Then, the restored image is obtained by regularizing the diffuse component only (see [4, 13] for details)

$$\min_{x \geq 0} D_{\mathcal{KL}}(g; Hx + b) + \beta\varphi_1(x_d). \tag{4}$$

The problems to which this strategy is applied are listed below. The evaluation of the models is done by computing the relative reconstruction error only on a window of interest of the diffuse component, located around the sources: this error is denoted by ρ_w.

Table 1. Comparison of the three models. In column–wise order: test problem, model, number of external iterations, total number of inner iterations, estimation of β, relative reconstruction error and time in secs. In the Bregman case, $\beta = 10\beta_{\text{opt}}$. The asterisk denotes that the maximum number of iterations allowed is reached.

Test problem	Model	k_{ext}	k_{tot}	β_k	err	time(s.)
cameraman	Model 1	8	815	$6.689 \cdot 10^{-3}$	$8.562 \cdot 10^{-2}$	16.67
	Model 2	451	–	$6.699 \cdot 10^{-3}$	$8.535 \cdot 10^{-2}$	30.28
	Bregman	6	3906	–	$8.730 \cdot 10^{-2}$	109.14
micro	Model 1	11	1458	$3.374 \cdot 10^{-3}$	$1.658 \cdot 10^{-1}$	9.77
	Model 2	*5000	–	$7.637 \cdot 10^{-3}$	$1.294 \cdot 10^{-1}$	94.41
	Bregman	9	4615	–	$8.370 \cdot 10^{-2}$	45.81
spacecraft	Model 1	41	3731	$1.000 \cdot 10^{-41}$	$1.000 \cdot 10^{0}$	258.73
	Model 2	*5000	–	$1.501 \cdot 10^{-4}$	$5.098 \cdot 10^{-1}$	360.20
	Bregman	9	27480	–	$3.780 \cdot 10^{-2}$	972.14

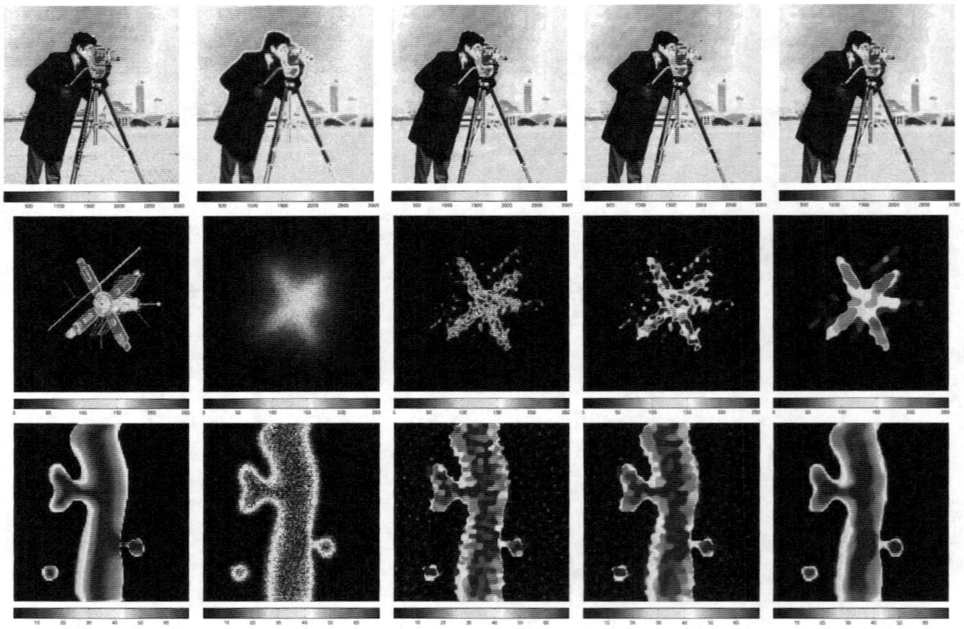

Figure 1. Visual inspection of the comparison. In column–wise order: original signal, blurred noisy images g, restored images by Model 1, Model 2 and inexact Bregman procedure, respectively.

- **YSO**: a 256×256 image of a single point source surrounded by a jet of circumstellar material. The imaging matrix is generated by means of `Software Package CAOS` [1], the relative ℓ_2 distance of g from x^* is 21.0254, b is $3.73 \cdot 10^3$. The source's intensity is $1.0184 \cdot 10^9$, while the maximum intensity on the x_d component is $2.3318 \cdot 10^6$ (see [4, 13]).
- **Binary Stars**: a 1024×1024 image with two point sources situated in a black disk at the centre of the diffuse component. H simulates the imaging matrix of a telescope. The ℓ_2 relative distance of g from x^* is 0.9823, b is 101.4. The point sources' intensities are $3.1411 \cdot 10^6$ while in the diffuse component the maximum value is $1.5200 \cdot 10^3$.

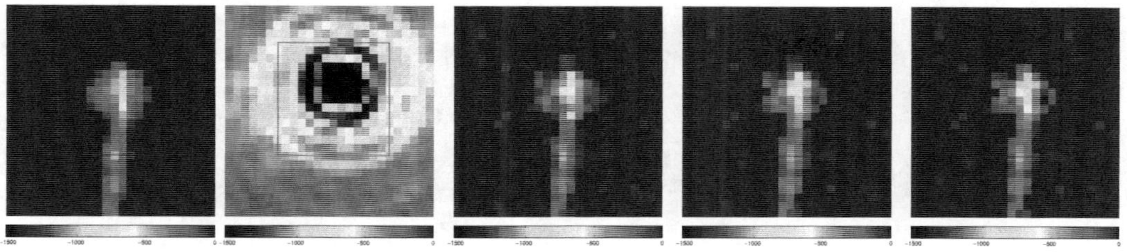

Figure 2. YSO reconstructions (particular). From left to right: x^*, g, Bregman procedure with Tikhonov regularization with $L = I$, $L = \nabla$ and $L = \nabla^2$ respectively. The highlighted square refers to the window on which ρ_w is computed. The images are displayed in a reverse square root scale.

For YSO problem, the Tikhonov regularization is used: $\varphi_1(x) = 1/2\|Lx\|_2$, with $L = I$ (identity matrix), $L = \nabla$ (discrete gradient) or $L = \nabla^2$ (discrete Laplacian). The optimal value for β is $1.6 \cdot 10^{-10}$, $5.0 \cdot 10^{-11}$ and $1.0 \cdot 10^{-11}$, respectively. In case of the Binary Stars, the **HS** regularization is used, β_{opt} is $5 \cdot 10^{-6}$. The solver used is SGP.

Table 2. YSO numerical results. φ_1 is the regularization function, β_{est} is the computed estimate, k_{ext} the number of external iterations, k_{tot} the total number of inner iterations and ρ_w is the relative reconstruction error of the window of interest. In the Bregman case, the β value denotes the value used in the procedure.

	Model 1				Inexact Bregman			
φ_1	β_{est}	k_{ext}	k_{tot}	ρ_w	β	k_{ext}	k_{tot}	ρ_w
$L = I$	$3.0 \cdot 10^{-9}$	11	6286	0.69	$2.0 \cdot 10^{-9}$	20	3590	0.3292
$L = \nabla$	$1.4 \cdot 10^{-9}$	10	5828	0.23	$0.4 \cdot 10^{-9}$	16	5287	0.3368
$L = \nabla^2$	$5.3 \cdot 10^{-10}$	9	4064	0.47	$3.3 \cdot 10^{-10}$	63	10071	0.2547

Table 2 shows the numerical results regarding the YSO problem. We observe that Model 1 can not achieve the optimal value for β, and does not provide reliable results. On the other hand, the inexact Bregman procedure allows to obtain satisfactory results even with a raw overestimation of the regularization parameter (Figure 2).

Finally, in Figure 3 the restoration of the image regarding the Binary Stars is presented. We

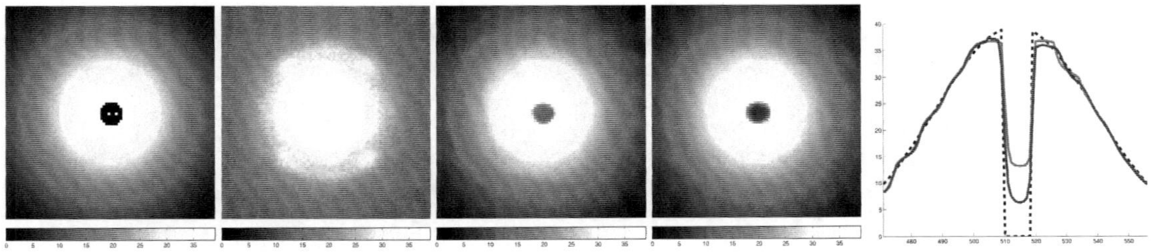

Figure 3. Binary star, window of interest. From left to right: x^*, g, restored image by SGP with β_{opt}, restored image at the 32th external iteration by Bregman procedure with $10\beta_{\text{opt}}$. The final panel is the superposition plot of the 513th line of x_d. The black dashed line is the original object, the red one refers to SGP reconstruction while the blue one refers to inexact Bregman reconstruction.

compare the results of Bregman procedure with the reconstruction obtained by solving (4) with

SGP method with $\beta = \beta_{\text{opt}}$. In such a way, the restored image shown in Figure 3 (3rd panel) is obtained in 178 iterations with a reconstruction relative error of 0.0941. Using the same method as inner solver in the inexact Bregman procedure, after 32 external iterations and 362 total inner iterations we obtain the reconstruction shown in Figure 3 (4th panel) where $\rho_w = 0.0822$: this result is achieved by setting $\beta = 10\beta_{\text{opt}}$. From the superposition plot of the line of diffuse data x_d related to the two stars (Figure 3, 5th panel), we can observe that the inexact Bregman procedure can reach the bottom of the central disk more precisely than the minimization of $D_{\mathcal{KL}} + \beta_{\text{opt}}\varphi_1$: this is due to the enhancing contrast feature of the inexact Bregman procedure.

4. Conclusions

In this work, the inexact Bregman procedure is compared with two different approaches for the estimate of the regularization parameter in presence of Poisson noise. It is shown that in case of high counts images the three models have a very similar behaviour, while for low counts images, and for HDR images, Model 1 and Model 2 do not provide satisfactory results. The Bregman procedure, instead, allows to treat such images and moreover permits to use an overestimation of the regularization parameter and at the same time to have a contrast enhancing.

References

[1] Arcidiacono C, Diolaiti E, Tordi M, Ragazzoni R, Farinato J, Vernet E and Marchetti E 2004 Layer-oriented simulation tool *Applied Optics IP* **43** 4288–4302
[2] Bardsley J M and Luttman A 2009 Total variation–penalized Poisson likelihood estimation for ill–posed problems *Adv. Comput. Math.* **31** 35–59
[3] Benfenati A and Ruggiero V 2013 Inexact Bregman iteration with an application to Poisson data reconstruction *Inverse Problems* **29**(6) 1–32
[4] Benfenati A and Ruggiero V 2015 Inexact Bregman iteration for deconvolution of superimposed extended and point sources *Communications in Nonlinear Science and Numerical Simulation* **20** 882–96
[5] Bertero M, Boccacci P, Desiderà G and Vicidomini G 2009 Image deblurring with Poisson data: from cells to galaxies *Inverse Problems* **25** 123006
[6] Bertero M, Boccacci P, Talenti G, Zanella R and Zanni L 2010 A discrepancy principle for Poisson data *Inverse Problems* **26** 10500
[7] Bonettini S, Zanella R and Zanni L 2009 A scaled gradient projection method for constrained image deblurring *Inverse Problems* **25** 015002
[8] Brune C, Sawatzky A and Burger M 2009 Bregman–TV–EM methods with application to optical nanoscopy, (Scale Space and Variationale Methods in Computer Vision) ed Tai X C et al Berlin : Springer 235–46
[9] Carlavan M and Blanc-Fèraud L 2012 Sparse Poisson noisy image deblurring *IEEE Trans. Image Process.* **21**(4) 1834-46
[10] De Mol C and Defrise M 2004 Inverse imaging with mixed penalties *Proceedings URSI EMTS 2004* (Ed. PLUS Univ. Pisa) 798–800
[11] Figueiredo M A T and Bioucas-Dias J M 2010 Restoration of Poissonian images using alternating direction optimization *IEEE Transactions on Image Processing* **19**(12) 3133–45
[12] Geman S and Geman D 1984 Stochastic relaxation, Gibbs distribution and the Bayesian restoration of images *IEEE Trans. Pattern Analysis and Machine Intelligence* **6** 721–41
[13] La Camera A, Antoniucci S, Bertero M, Boccacci P, Lorenzetti D, Nisini B and Arcidiacono C 2014 Reconstruction of high dynamic range images: simulations of LBT observations of a stellar jet, a pathfinder study for future AO–assisted giant telescopes *Publications Astron. Soc. Pacific* **126** 180–93
[14] Osher S, Burger M, Goldfarb D, Xu J and Yin W 2005 An iterative regularization method for total variation-based image restoration *SIAM Journal on Multiscale Modeling and Simulation* **4**(2) 460–89
[15] Teuber T Steidl G and Chan R H 2013 Minimization and parameter estimation for seminorm regularization models with I-divergence constraints *Inverse Problems* **29** 035007
[16] Willett R M and Nowak R D 2003 Platelets: A multiscale approach for recovering edges and surfaces in photon limited medical imaging *IEEE Transactions on Medical Imaging* **22** 332–50
[17] Zanella R, Boccacci P, Zanni L and Bertero M 2009 Efficient gradient projection methods for edge-preserving removal of Poisson noise *Inverse Problems* **25** 045010
[18] Zanni L, Benfenati A, Bertero M and Ruggiero V 2014 Numerical methods for parameter estimation in Poisson data inversion *Journal of Mathematical Imaging and Vision* Springer US

Features selection and classification to estimate elbow movements

A Rubiano[1,3], J L Ramírez[1], M N El Korso[1], N Jouandeau[2], L Gallimard[1] and O Polit[1]

[1]LEME, Université Paris Ouest Nanterre la Défense, 50 rue de Sèvres, 92410 Ville d'Avray, FR
[2]LIASD, Université Paris 8,2 Rue de la Libert, 93526 Saint-Denis, FR
[3]Universidad Militar Nueva Granada, Cr 11 101-80, 110111 Bogotá, COL

E-mail: a.rubiano_fonseca@u-paris10.fr

Abstract. In this paper, we propose a novel method to estimate the elbow motion, through the features extracted from electromyography (EMG) signals. The features values are normalized and then compared to identify potential relationships between the EMG signal and the kinematic information as angle and angular velocity. We propose and implement a method to select the best set of features, maximizing the distance between the features that correspond to flexion and extension movements. Finally, we test the selected features as inputs to a non-linear support vector machine in the presence of non-idealistic conditions, obtaining an accuracy of 99.79% in the motion estimation results.

1. Introduction

The Electromyography (EMG) signal is a measure of the electrical activity produced by muscles during the movement. Therefore, it is possible to estimate the motion through the analysis of these signals [1]. Even that, due to the nature of the EMG the estimation is not straightforward [2].

In the literature, several relevant studies have been carried out. Some of them are focused on the relation of the EMG signals and the movement, and some others in the classification of signals to estimate movements. Regarding the approach to the classification of EMG signals, Oskoei and Huosheng [3] performed the classification of five movements of the hand, they reached an accuracy of 94%. Likewise, in [4], the authors proposed the identification of the wrist and the ring finger movements, and they earned an accuracy of 87.3%. In the same way, Wang *et al.* [5] put forward the recognition of eight grasping gestures and achieved an accuracy between 96.9% and 99.65%. Concerning the analysis of the relationship between EMG and motion, Kamavuako *et al.* [6] carried out a study to find the relationship between the grasping force and the EMG signals.

Nevertheless, these important studies have not established a clear relationship, and so, the classification process could be improved. Consequently, we introduce a new method to analyse the EMG signals based on features. A feature is a scalar value extracted from the EMG signal in time or frequency domains that contains relevant information that could be linked with motion. Our method, analyse the kinematic parameters as angular velocity an angular position, to extract in an accurate way the signal fragment that matches exactly with the appropriate movement. Therefore, we enhance the possibility of establishing a relationship EMG-motion to improve the process of motion estimation.

Afterwards, we perform a correlation between the kinematic data and the EMG signals. This relationship allows to propose a criterion for selecting the couple of feature that best separates the flexion and extension movements.

Finally, the results of our criterion (best features) are validated using a non-linear SVM algorithm (which is optimized with respect to the so-called overfitting and radius coefficients). We add simulated white zero-mean Gaussian noise to the signals for assessing our method in non-idealistic condition. As a result, we achieved a classification percentage of 99.79% which is higher than the classification achieved in the state of the art.

2. Experimental Set-up

In this study, we consider three healthy subjects that execute flexion and extension movements, the first, second and third subject execute one, four and six trials, respectively. For each movement of flexion and extension, we measured EMG signals of the triceps and the biceps (due to their high electrical activity during these movements [7]) and tracked the upper limb motion.

The motion was tracked using seven 3D VICON cameras with a sampling frequency of 200 Hz, and 30 retro-reflective markers distributed as shown in Fig. 1. The EMG signals were captured using wireless sensors with a sampling frequency of 1 kHz.

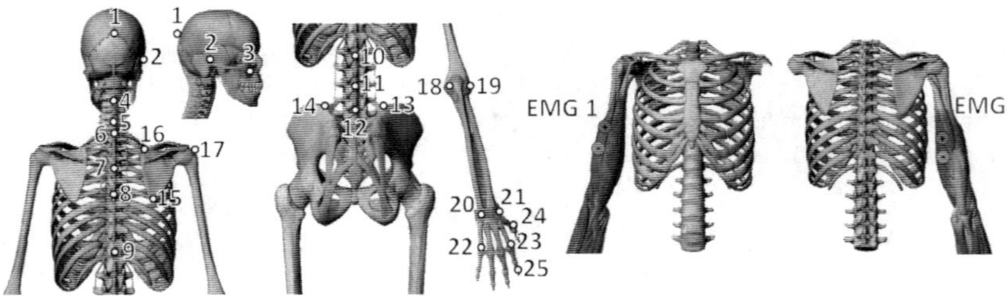

Figure 1. The Position of retro-reflective markers and EMG sensors.

To preprocess the experimental data in an accurate way, we develop an open source software[1] that allows us to display the measures (motion and EMG) and calculate kinematic parameters as rotation and angular velocity. Our software synchronizes motion capture and EMG signals (measured in real time) to perform an accurate analysis of the relation between motion and EMG.

3. Kinematic analysis and features selection

In this section, we present the kinematic analysis of the upper limb, and we introduce the relation between the EMG signals and kinematic parameters. afterwards, we perform a feature extraction and we propose a new strategy to select the best couple of features.

3.1. Kinematic analysis of the movement

The synchronization of the 3D motion and EMG signals, allows to evaluate the kinematics data i.e. the elbow angle, θ, and angular velocity η. The Fig. 2 shows these values for the eleven trials. To calculate the angle, we introduce a vector linking the elbow to the wrist (forearm's vector) and a vector linking the elbow to the shoulder (arm's vector). The arm's vector $\boldsymbol{x}_a$, is defined from the point p_1 (located at the middle point between markers 18 and 19 placed over

[1] The open source software is available at `http://leme.u-paris10.fr/promain-565200.kjsp`

the coronoid process of the ulna and the radius) to p_3 (marker 17 placed over the epicondyle of the humerus). The forearm vector, $\boldsymbol{x}_{fa}$, is defined from p_1 to p_2 (located at the middle point between markers 20 and 21 placed over the styloid process of the ulna and the radius). Consequently, the elbow angle is evaluated by $\theta = \arccos\left(\langle \boldsymbol{x}_a, \boldsymbol{x}_{fa}\rangle / \|\boldsymbol{x}_a\|\|\boldsymbol{x}_{fa}\|\right)$. The sign of the angular velocity, calculated as $\eta = d\theta/dt$, defines if the movement corresponds to an extension (positive) or a flexion (negative). Considering that the motion capture and the EMG signals are synchronized, we use the sign of the angular velocity η to extract the part of the EMG signal that corresponds to a particular movement (flexion or extension).

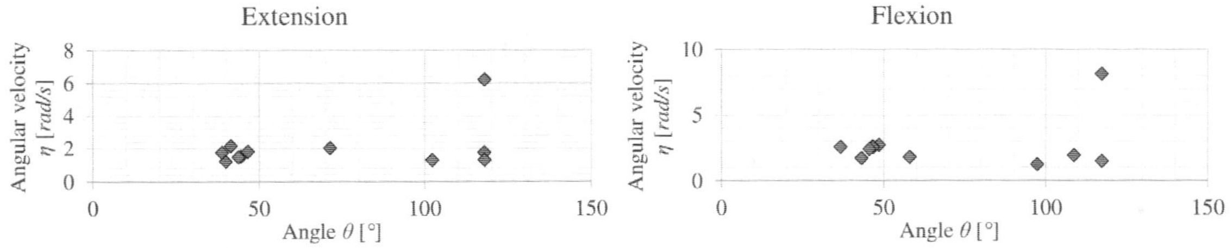

Figure 2. Extracted kinematical parameters after the pre-processing step.

3.2. Relationship between EMG and Kinematic

The EMG signal is a measure of the electrical activity produced by muscle contractions during motion. From these signals, it is possible to extract scalar values that could contain relevant information of the movement, and so on, establish a relationship between EMG and motion. Consequently, we choose the following set of features in time and frequency domain:

- Mean Absolute Value: $\mathrm{MAV}(\boldsymbol{y}) = n^{-1}\sum_{i=1}^{n} |[\boldsymbol{y}]_i|$, where $[\boldsymbol{y}]_i$ represents the i-th sample of the signal $\boldsymbol{y}$, and n is the number of samples.
- Mean value: $\mathrm{M}(\boldsymbol{y}) = n^{-1}\sum_{i=1}^{n}[\boldsymbol{y}]_i$.
- Entropy: $\mathrm{Ent}(\boldsymbol{y}) = -\sum_i e_i^2 \log_2(e_i^2)$ where e_i represents the projection coefficients of the signal $\boldsymbol{y}$ in an orthonormal basis [8].
- Harmonic mean: $\mathrm{HM}(\boldsymbol{y}) = n(\sum_{i=1}^{n}[\boldsymbol{y}]_i^{-1})^{-1}$.
- Mean frequency: $\mathrm{MF}(\boldsymbol{y}) = (\sum_{j=1}^{N} I_j)^{-1}\sum_{j=1}^{N} I_j f_j$ where N denotes the number of harmonics in the spectrum, I_j represents the magnitude of the j-th harmonic, and f_j is the frequency of the j-th harmonic.

Once we apply the feature extraction models to all trials, we get a set of scalar values that correspond to the flexion and the extension. Considering that the values have different scales, we apply a normalisation. The table 1 shows the normalised values of the features; the row one contains the values for trial one, the row two represents trial two, and so on.

Comparing the normalised values of the features during the different movements, we find that M and HM have significant variations and few local minimal and maximal values (see figure 3 left side). These long fluctuations could be useful to model small angle or velocity changes. But in the opposite, the identification of the complete range will be more difficult. Moreover, Ent, MAV and MF show a similar pattern, a more regular behaviour and several local minimal and maximal values (see figure 3 right). These particularities make possible to model and consequently describes a long range movement. It shows that there is a relation between the features extracted from the EMG signals and the motion. Thus we propose a strategy to choose the best features to identify the flexion and extension movements.

Table 1. The normalised values of the features.

Flexion										Extension									
Biceps					Triceps					Biceps					Triceps				
Ent	Mav	M	HM	MF	Ent	Mav	M	HM	MF	Ent	Mav	M	HM	MF	Ent	Mav	M	HM	MF
−0.47	−0.57	0.01	−0.31	−0.69	−0.07	−0.94	1.36	−2.50	−0.66	2.25	2.81	−1.60	3.02	2.14	2.55	2.90	−0.57	0.45	1.65
−0.26	−0.49	−0.46	−0.31	−0.42	−0.57	−0.82	0.03	0.52	−0.88	−1.05	−0.66	0.52	−0.30	−1.00	−0.86	−0.62	−0.40	0.45	−1.10
−0.03	−0.34	−0.37	−0.31	−0.19	−0.52	−0.80	0.03	0.53	−0.86	−0.92	−0.61	0.47	−0.30	−0.79	−0.74	−0.56	−0.73	0.45	−1.09
−0.63	−0.03	−1.55	−0.32	0.28	0.47	0.07	1.28	−0.87	−0.31	0.19	−0.01	−1.06	−0.30	0.04	0.16	−0.04	−0.34	−2.02	0.07
0.00	−0.42	−0.41	2.43	−0.55	−0.01	−0.59	0.30	0.53	−0.09	−0.76	−0.59	0.74	−0.30	−1.28	−1.20	−0.81	0.04	0.45	−1.71
2.03	2.20	0.41	−0.31	2.67	1.38	1.32	−0.33	0.53	1.49	0.91	0.34	−0.81	−0.30	1.16	0.45	−0.03	2.61	0.45	0.34
−1.02	−0.89	0.33	−0.31	−0.57	−0.74	−0.11	−0.48	0.53	0.66	−0.51	−0.42	0.80	−0.30	−0.12	−0.19	−0.14	0.15	0.45	1.12
−0.63	−0.03	−1.55	−0.32	0.28	0.47	0.07	1.28	−0.87	−0.31	0.19	−0.01	−1.06	−0.30	0.04	0.16	−0.04	−0.34	−2.02	0.07
−0.35	−0.56	1.40	−0.31	−0.64	−1.80	−0.71	−0.73	0.54	−1.25	−0.68	−0.57	0.44	−0.30	−0.71	−0.19	−0.27	0.27	0.45	0.50
1.84	1.69	1.03	1.32	0.62	1.77	2.27	−1.62	0.53	1.89	0.77	0.23	−0.07	−0.30	0.52	0.42	−0.02	0.56	0.44	0.47
−0.48	−0.58	1.17	−1.24	−0.77	−0.38	0.22	−1.12	0.53	0.31	−0.39	−0.52	1.64	−0.30	0.01	−0.54	−0.35	−1.25	0.45	−0.32

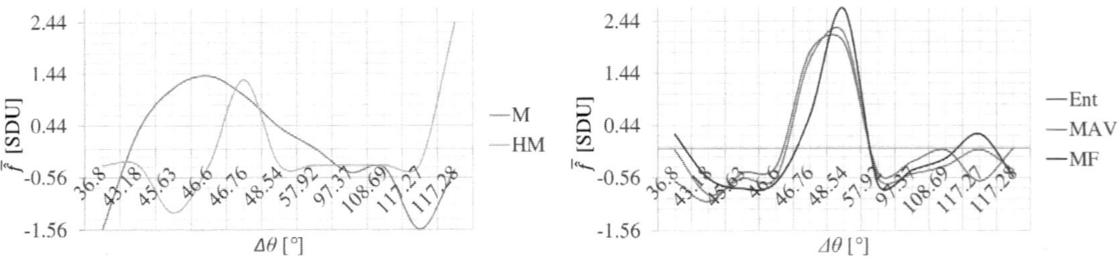

Figure 3. The behaviour of the features with respect to the movement of flexion.

3.3. Features selection

For this purpose, each movement is represented in an n-dimensional space Υ. For two muscles, we extract two features from the biceps signal and one for the triceps, and the space Υ will have four dimensions. One point in this space represents one movement, flexion or extension.

The strategy for selecting the best couple of features is based on the maximization of the Euclidean distance between flexion and extension points. For a couple of features $(\hat{f}_1, \hat{f}_2)$, for $\hat{f}_1 \neq \hat{f}_2$, this function satisfies:

$$(\hat{f}_1, \hat{f}_2) = \underset{f_1, f_2}{\arg\max} \left(\min \left(\sum_{i=1}^{2} \left(\mathcal{F}_{f_i}^{\text{ext}} - \mathcal{F}_{f_i}^{\text{fle}} \right)^2 \right) \right) \tag{1}$$

where $\mathcal{F}_{f_i}^{\text{ext}}$ and $\mathcal{F}_{f_i}^{\text{fle}}$ denote the values of the ith feature for extension and flexion, respectively. As a result of the proposed criterion (1), the couple (Ent, MF) shows the best fitness among all the features introduced above.

4. Supervised Learning Classification

The supervised learning method is introduce in this section. It is used to assess the performance of the selected features for motion classification. Furthermore, external perturbations [2] of the EMG signal is simulated with additive white zero-mean Gaussian noise (AWZMGN) [1].

4.1. Support Vector Machine

Support vector machine (SVM) is a supervised learning method that can separate a set of samples $\boldsymbol{F}$ into two categories $z_l \in \{-1, 1\}$. The method separates the categories using a subset $\boldsymbol{F}'$ as

[2] Sweat and fatigue of the subject, displacement of the recording electrodes, thermal disturbances, among others.

training data to calculate an optimal hyperplane between them. The subset $\boldsymbol{F}'$ is composed by q samples denominated support vectors $\boldsymbol{f}'_q$.

In this study, the set of samples $\boldsymbol{F}$ is composed of the selected features (MF and Ent) extracted from the biceps' and the triceps' EMG signals. Consequently, each member of $\boldsymbol{F}$ is defined as $[\boldsymbol{F}]_l = \boldsymbol{f}_l = [\text{Ent}(\boldsymbol{y}_{bi}), \text{MF}(\boldsymbol{y}_{bi}), \text{Ent}(\boldsymbol{y}_{tr}), \text{MF}(\boldsymbol{y}_{tr})]$ where index l is the trial number. Furthermore, the $\boldsymbol{F}'$ is composed by three elements of $\boldsymbol{F}$. Considering that a linear separation is not possible, we propose the use of a non-linear SVM classifier, in which the optimal hyperplane is obtained by solving the following quadratic programming problem [9]:

$$\min_{\boldsymbol{h},b,\boldsymbol{\kappa}} \frac{1}{2}\boldsymbol{h}^T\boldsymbol{h}+C\sum_{q=1}^{m}[\boldsymbol{\kappa}]_q \quad \text{s.t.} \quad z_q\left(\boldsymbol{h}^T\Theta(\boldsymbol{f}'_q)+\gamma\right) \geq 1-[\boldsymbol{\kappa}]_q \quad \text{and} \quad [\boldsymbol{\kappa}]_q \geq 0 \quad \text{for} \quad q=1,\ldots,m \tag{2}$$

where $[\boldsymbol{\kappa}]_q$ is the error soft margin, $\boldsymbol{h}$ and γ determine the hyperplane in feature space. C denotes a term to control the overfitting, m represents the amount of support vectors inside $\boldsymbol{F}'$, and Θ maps $\boldsymbol{f}'_q$ into a higher-dimensional space. The following decision function represents the solution of the following problem:

$$\Psi(\boldsymbol{f}_l) = sign\left(\sum_{q=1}^{m}\beta_q z_q \mathcal{K}(\boldsymbol{f}'_q, \boldsymbol{f}_l)+\gamma\right) \quad \text{for} \quad \mathcal{K}(\boldsymbol{f}'_q, \boldsymbol{f}_l) = \exp^{-\frac{(\boldsymbol{f}'_q-\boldsymbol{f}_l)(\boldsymbol{f}'_q-\boldsymbol{f}_l)^T}{2\sigma^2}} \tag{3}$$

where β_q denotes the Lagrange coefficients [8], $\mathcal{K}(\boldsymbol{f}'_q, \boldsymbol{f}_l)$ is the radial basis kernel function[3] and σ is a positive parameter associated with the radius.

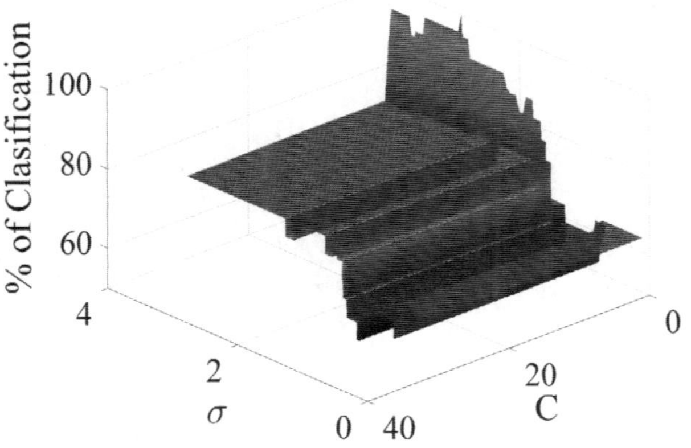

Figure 4. Percentage of correct classification as function of σ and C.

Subsequently, we apply the decision function $\Psi(\boldsymbol{f}_l)$ to the elements of $\boldsymbol{F}$ that have not been used as support vectors, and we compare the result with the corresponding z_l which are known. If the result of the decision function match with z_l, then the classification is correct.

4.2. Classifier Implementation

It is necessary to take into account the influence of the parameters σ and C for the classification performance. Therefore, we propose to find the optimal values σ_{opt} and C_{opt}, following the

[3] This function allows classify using a Euclidean distance.

criterion:

$$(C_{\text{opt}}, \sigma_{\text{opt}}) = \arg\max_{C,\sigma} \sum_{l=1}^{l} \delta(\Psi(\boldsymbol{f}_l) - z_l) \tag{4}$$

where $\delta(.)$ is the Dirac function (equals to one if its argument is zero and equal to zero elsewhere).

We implement the SVM defined by equations (2) and (3), using three support vectors, then the criterion (4) is applied. We find that the optimal value is not unique. Fig. 4 shows the relation between C, σ and the percentage of classification; the red zone is the optimal zone where the classification percentage reach 100%. Therefore, it must be denoted that the classifier identifies the movements of the first subject although the support vectors is based on features extracted from the signal of subjects two and three; the test is performed with all subjects.

Finally, we add AWZMGN to the original set of EMG signals using 1000 Monte Carlo trials for different values of signal to noise ratio (SNR). Afterwards, we use the noised signals as the new set of samples $\boldsymbol{F}$ and we perform the SVM classification described in section 4.1. We achieve a classification percentage of 99.79% for an SNR of $37.5dB$.

5. Conclusions

In this paper, the relationship between EMG signals from the biceps and triceps for flexion and extension movements of the elbow is analysed. A novel method to assess and select the best features is presented for the classification of the movements. A criterion is defined to select the best features associated with the motion.

The test performed with the non-linear SVM algorithm, using only 13.63% of the whole data as training data, allow us to achieve a 100% successful classification. For a new subject and after adding simulated AWZMGN, the classification reach a percentage of 99.79%. Therefore, the proposed method to analyse EMG signals is efficient and robust.

Acknowledgments

We express our gratitude to the Université Paris Lumières for the financial support through the project PROMAIN, and to the enterprise COGITOBIO where the experiments have been held.

References

[1] Reaz, M. B. I. and Hussain, M. S. and Mohd-Yasin, F., *Techniques of EMG signal analysis: detection, processing, classification and applications*, Biological Procedures Online, Vol. 8, 2006, pp. 11-35.

[2] Alkan, A. and Gnay, M., *Identification of EMG signals using discriminant analysis and SVM classifier*, Expert Systems with Applications, Vol. 39, 2012, pp. 44-47.

[3] Oskoei, M.A. and Huosheng, H. *Support Vector Machine-Based Classification Scheme for Myoelectric Control Applied to Upper Limb*, IEEE Transactions on Biomedical Engineering, Vol. 55 2008, pp. 1956-1965.

[4] Naik, G. R. and Kumar, D.K. and Jayadeva, J., *Twin SVM for Gesture Classification Using the Surface Electromyogram,* IEEE Transactions on Information Technology in Biomedicine, Vol. 14, 2010, pp. 301-308.

[5] Wang, N. and Chen, Y. and Zhang, X., *The recognition of multi-finger prehensile postures using LDA*, Biomedical Signal Processing and Control, Vol. 8 2013, pp. 706-712.

[6] Kamavuako, E. N. and Farina, D. and Yoshida, K. and Jensen, W., *Relationship between grasping force and features of single-channel intramuscular EMG signals*, Journal of Neuroscience Methods, Vol. 185, 2009, pp 143-150.

[7] Ali, A. and Sundaraj, K. and Ahmad, R. B. and Ahamed, N. U. and Islam, A., *Surface electromyography for assessing triceps brachii muscle activities: A literature review,* Biocybernetics and Biomedical Engineering, Vol. 33, 2013, pp. 187-195.

[8] Liu, Q. and Zhao, X. and Hou, Z., *Metric learning for event-related potential component classification in EEG signals*, Proceedings of the 22nd European Signal Processing Conference (EUSIPCO), 2014, Lisbon, Portugal, pp. 2005-2009.

[9] Chamasemani, F. F. and Singh, Y. P., *Multi-class Support Vector Machine (SVM) Classifiers - An Application in Hypothyroid Detection and Classification*, Sixth conference on Bio-Inspired Computing: Theroies and applications, IEEE, 2011, Penang, Malaysia, pp. 351-356.

Journal of Physics: Conference Series **657** (2015) 012013 doi:10.1088/1742-6596/657/1/012013

Asymptotic of Sparse Support Recovery for Positive Measures

Q Denoyelle, V Duval and G Peyré,

CEREMADE, Université Paris-Dauphine
INRIA
CEREMADE, CNRS, Université Paris-Dauphine

E-mail: `denoyelle@ceremade.dauphine.fr`, `vincent.duval@inria.fr`, `gabriel.peyre@ceremade.dauphine.fr`

Abstract. We study sparse spikes deconvolution over the space of Radon measures when the input measure is a finite sum of positive Dirac masses using the BLASSO convex program. We focus on the recovery properties of the support and the amplitudes of the initial measure in the presence of noise when the minimum separation distance t of the input measure (the minimum distance between two spikes) tends to zero. We show that when $\|w\|_2/\lambda$, $\|w\|_2/t^{2N-1}$ and λ/t^{2N-1} are small enough (where λ is the regularization parameter, w the noise and N the number of spikes), which corresponds roughly to a sufficient signal-to-noise ratio and a noise level and a regularization parameter small enough with respect to the minimum separation distance, there exists a unique solution to the BLASSO program with exactly the same number of spikes as the original measure. We provide an upper bound on the error with respect to the initial measure. As a by-product, we show that the amplitudes and positions of the spikes of the solution both converge towards those of the input measure when λ and w drop to zero faster than t^{2N-1}.

1. Introduction – Sparse Spikes Deconvolution

Sparse spikes deconvolution aims at recovering a sparse spikes train $m_{a_0,tz_0} \stackrel{\text{def.}}{=} \sum_{i=1}^N a_{0,i}\delta_{tz_{0,i}}$ with amplitudes $a_{0,i} \in \mathbb{R}$ and positions $tz_{0,i}$ on the torus $\mathbb{T}$, from low resolution observations $y_t + w$ where $y_t \stackrel{\text{def.}}{=} \Phi m_{a_0,tz_0}$. Here $\Phi m \stackrel{\text{def.}}{=} \varphi \star m$ is a low pass filter against an arbitrary smooth kernel φ and w is the noise contaminating the observations. This problem is a toy model for many applications, ranging from seismic trace deconvolution to stars field observation in astrophysics and spike sorting in neuro-imaging. The parameter $t > 0$ controls the minimum separation distance between the spikes, and we study the case when t is small. Note that while we expose the problem in 1-D, our analysis carries over in arbitrary dimension.

The recovery is obtained by the BLASSO convex optimization problem over the space $\mathcal{M}$ of Radon measure on $\mathbb{T}$:

$$\min_{m\in\mathcal{M}} \frac{1}{2}\|\Phi m - (y_t + w)\|^2 + \lambda|m|(\mathbb{T}). \qquad (\mathcal{P}_\lambda(y_t + w))$$

where $\lambda > 0$ is the regularization parameter and $|m|(\mathbb{T})$ is the total variation of a measure m on the torus (not to be mistaken for the total variation of an image), that extends the discrete ℓ^1 norm of vectors to measures:

$$|m|(\mathbb{T}) \stackrel{\text{def.}}{=} \sup\left\{\int_{\mathbb{T}} \psi \mathrm{d}m;\ \psi \in C(\mathbb{T}), \|\psi\|_\infty \leq 1\right\}.$$

Without the presence of noise on the data $y_t = \Phi m_{a_0,tz_0}$, we rather consider the Basis Pursuit for measures ([11, 3]):

$$\min_{m\in\mathcal{M},\ \Phi m = y_t} |m|(\mathbb{T}). \qquad (\mathcal{P}_0(y_t))$$

We refer the reader to [1, 4, 6, 7, 8, 5] for more details about the BLASSO program, the Basis Pursuit for measures and algorithms to compute numerically the solution to this problem.

The recovery performance of $(\mathcal{P}_0(y_t))$ and $(\mathcal{P}_\lambda(y_t + w))$ (i.e. how m is close to m_{a_0,tz_0}) for arbitrary a_0 has been the focus of several recent contributions. The breakthrough paper [2] shows that stable recovery of m_{a_0,tz_0} holds provided that the spikes are sufficiently separated, i.e. when t is large enough. Subsequent contributions analyzed carefully this stability in terms of recovered spikes concentration [6, 7] and exact support recovery [5].

For positive spikes (i.e. $a_{0,i} > 0$), the picture is radically different, since exact recovery without noise (i.e. $(w, \lambda) = (0, 0)$) holds for all $t > 0$ in the case of the ideal low-pass filter, see for instance [4]. Stability constants however explode as $t \to 0$. A recent work [8] shows however that stable recovery is obtained if the signal-to-noise ratio grows faster than $O(1/t^{2N})$, closely matching optimal lower bounds of $O(1/t^{2N-1})$ obtained by combinatorial methods, as also proved recently [10]. Our main contribution is to show that the same exact $O(1/t^{2N-1})$ signal-to-noise scaling in fact guarantees a perfect support recovery of the spikes under a certain non-degeneracy condition on the filter. This extends, for positive measures, the initial results of [5] by providing an asymptotic analysis when $t \to 0$.

2. Asymptotic of Exact Support Recovery

2.1. Dual Certificate

An approach to study the problem $\mathcal{P}_\lambda(y_t + w)$ is to apply Fermat's rule. Indeed, a discrete measure $m_{a,tz} \stackrel{\text{def.}}{=} \sum_{i=1}^N a_i \delta_{tz_i}$ is solution of $\mathcal{P}_\lambda(y_t + w)$ if and only if there exists $\eta \in \mathcal{C}(\mathbb{T})$ such that:

$$\Phi^*(\Phi m_{a,tz} - y_t) + \lambda\eta = 0 \quad \text{and} \quad \eta \in \partial|m_{a,tz}|(\mathbb{T})$$

where $\partial|m_{a,tz}|(\mathbb{T})$ is the subdifferential of $|\cdot|(\mathbb{T})$ at $m_{a,tz}$ and is given by:

$$\partial|m_{a,tz}|(\mathbb{T}) = \{\eta \in \mathcal{C}(\mathbb{T});\ \|\eta\|_\infty \leq 1 \text{ and } \forall i = 1, \ldots, n,\ \eta(tz_i) = \operatorname{sign}(a_i)\}.$$

In that case, η is what we called a *certificate* for $\mathcal{P}_\lambda(y_t + w)$ of $m_{a,tz}$.

Similarly, finding a continuous function η verifying:

$$\eta \in \operatorname{Im}\Phi^* \quad \text{with} \quad \eta \in \partial|m_{a_0,tz_0}|(\mathbb{T}),$$

assures that m_{a_0,tz_0} is a solution of $\mathcal{P}_0(y_t)$. η is then a certificate for $\mathcal{P}_0(y_t)$ of m_{a_0,tz_0}. It interpolates the sign of the amplitudes at the spikes' positions and is lower than or equal to 1 in uniform norm.

2.2. Construction of a Solution to $\mathcal{P}_\lambda(y_t + w)$

Our approach to study $\mathcal{P}_\lambda(y_t+w)$ is to proceed in a constructive manner and then find sufficient conditions on the parameters w, λ, t (typically a low noise regime with the right behavior of w and λ with respect to t) to assure that the unique solution of $\mathcal{P}_\lambda(y_t + w)$ is composed of N spikes. With that knowledge we can then provide a study of the properties of this solution (for example how far are the spikes from the original measures with respect to noise level).

So the idea is to build a candidate solution $m_{a,tz}$ and then verify that $\eta_{\lambda,t} = \frac{1}{\lambda}\Phi^*(y_t + w - \Phi m_{a,tz})$ is a certificate, for any (w, λ) close from $(0, 0)$ with the right behavior with respect to t. We do not provide the proofs of the results detailed below due to a lack of space. Interested readers may refer to [9].

For the first step we consider the measures $m_{a,tz}$ composed of N spikes that are solutions of the following two equations:

$$\forall i = 1, \ldots, N,\ \Phi^*_{tz_i}(\Phi m_{a,tz} - (y_t + w)) + \lambda = 0 \quad \text{and} \quad (\Phi^*_{tz_i})'(\Phi m_{a,tz} - (y_t + w)) = 0.$$

These equations mean respectively that $\eta_{\lambda,t}(tz_i) = 1$ and $\eta'_{\lambda,t}(tz_i) = 0$ for all i i.e. $\eta_{\lambda,t}$ interpolates the sign of the amplitudes with zero derivatives. By applying the Implicit Function Theorem with respect to (λ, w) around $(0, 0)$, we obtain a local parametrization of the measures $m_{a,tz}$ solutions of the equations. Now that we have a candidate solution $m_{a,tz}$ for any (λ, w) around $(0, 0)$ and $t > 0$, we can move to the second step where it remains to show that $\|\eta_{\lambda,t}\|_\infty \leq 1$ provided sufficient hypothesis on (λ, w) with respect to t.

Following [5], we introduce below the so called "vanishing derivatives pre-certificate" $\eta_{V,t}$, which is a function defined on $\mathbb{T}$ that interpolates the spikes positions and signs (here +1). Note that $\eta_{V,t}$ can be computed in closed form by solving a linear system [5].

Proposition 1 (Vanishing derivatives pre-certificate). *One has the uniform convergence $\eta_{\lambda,t} \to \eta_{V,t}$ as $\frac{w}{\lambda} \to 0$ and $\lambda \to 0$ where $\eta_{V,t} = \Phi^* p_{V,t}$ with*

$$p_{V,t} = \underset{p \in L^2(\mathbb{T})}{\operatorname{argmin}} \left\{ \|p\|_2;\ (\Phi^* p)(tz_i) = 1 \text{ and } (\Phi^* p)'(tz_i) = 0\ \forall i \right\}.$$

Recall the conditions that define a certificate for $\mathcal{P}_0(y_t)$ that we saw in the previous section, then note that $\eta_{V,t}$ is a good candidate of being one but is not in general (hence its name). It is one if and only if $\|\eta_{V,t}\|_\infty \leq 1$, and then m_{a_0,tz_0} is recovered by the Basis Pursuit $(\mathcal{P}_0(y_t))$.

Definition 1 (Non-degeneracy of $\eta_{V,t}$). *We say that $\eta_{V,t}$ is non-degenerate if $|\eta_{V,t}| < 1$ outside the position $tz_{0,i}$ and $\eta''_{V,t}(tz_{0,i}) \neq 0$ for $i = 1, \ldots, N$.*

It is shown in [5] that $\eta_{V,t}$ being non-degenerate implies that solving $(\mathcal{P}_\lambda(y_t + w))$ in a certain low noise regime ($w/\lambda \to 0$ and $\lambda \to 0$) gives a unique solution with the correct number N of spikes, i.e. one obtains exact support recovery.

Proposition 1 is a step forward towards the proof of $\eta_{\lambda,t}$ being a certificate. Indeed if we are able to prove that $\eta_{V,t}$ is non-degenerate for all $t > 0$ then we can prove the expected result in a sufficient low noise regime with respect to t. So now we investigate the problem when $t \to 0$ by describing how $\eta_{V,t}$ behaves.

Figure 1, top row, shows graphically that $\eta_{V,t}$ converges with $t \to 0$ toward a fixed function η_W that does not depend on z_0 and t. This function is defined in the following proposition, which ensures that the non-degeneracy of $\eta_{V,t}$ can thus be checked independently of z_0 and t provided that t is small.

Proposition 2. *One has the uniform convergence $\eta_{V,t} \to \eta_W = \Phi^* p_W$ as $t \to 0$, where*

$$p_W = \underset{p \in L^2(\mathbb{T})}{\operatorname{argmin}} \left\{ \|p\|_2; (\Phi^* p)(0) = 1, (\Phi^* p)^{(i)}(0) = 0\ \forall i \leq 2N - 1 \right\}.$$

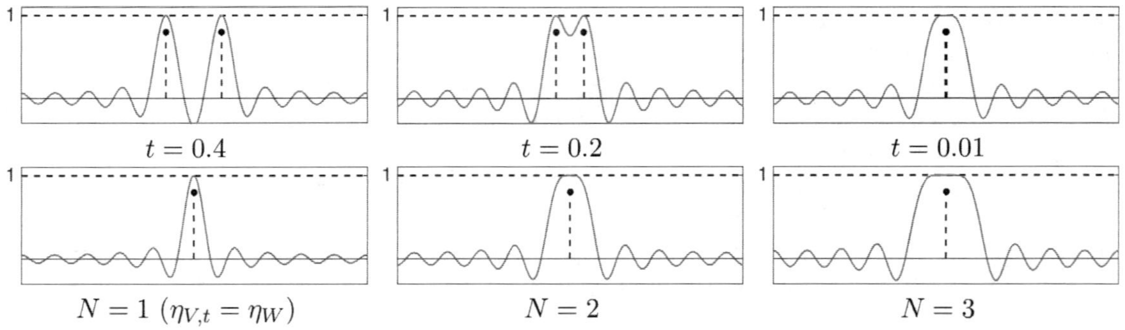

Figure 1. *Top row:* $\eta_{V,t}$ for several values of t, showing the convergence toward η_W. The filter φ is an ideal low-pass filter with a frequency cutoff $f_c = 10$. *Bottom row:* η_W for several value of N.

Furthermore, if $|\eta_W| < 1$ *outside* 0 *and* $\eta_W^{(2N)}(0) \neq 0$ *(we then say that* η_W *is* $(2N-1)$*-non-degenerate), then* $\eta_{V,t}$ *is non-degenerate for small* t*.*

Figure 1, bottom row, shows η_W for several values of N. Note how it becomes flatter at 0 as N increases. This implies that $\eta_{V,t}$ for small t gets closer to degeneracy as N increases. This is reflected in our main contribution (Theorem 1 below) where signal-to-noise ratio is required to scale like $1/t^{2N-1}$.

So one can sum up the convergence of the pre-certificates as follows:

$$\eta_{\lambda,t} \underset{\frac{w}{\lambda}\to 0, \lambda\to 0}{\longrightarrow} \eta_{V,t} \underset{t\to 0}{\longrightarrow} \eta_W.$$

As a result, we can state as in Proposition 2 that if η_W is $(2N-1)$-non-degenerate, then $\eta_{\lambda,t}$ is a certificate. In fact the next Proposition shows that $\eta_{\lambda,t}$ verifies the same non-degeneracy properties as $\eta_{V,t}$ (which implies in particular that $\eta_{\lambda,t}$ is certificate).

Proposition 3. *If* η_W *is* $(2N-1)$*-non-degenerate then for* $\frac{w}{\lambda}$*,* λ *and* t *small enough, one has* $|\eta_{\lambda,t}| < 1$ *outside the positions* tz_i *and* $\eta''_{\lambda,t}(tz_i) \neq 0$*.*

Remark 1. η_W *being* $(2N-1)$*-non-degenerate depends only on the filter* φ*. See Figure 2 for a* $(2N-1)$*-non-degenerate* η_W *(when* φ *is the Dirichlet kernel) and a degenerate* η_W *(a low-pass filter with high extreme Fourier coefficients). However* $\eta_W^{(2N)}(0) \neq 0$ *is true when the family* $(\varphi, \varphi', \ldots, \varphi^{(2N)})$ *is linearly independent in* $L^2(\mathbb{T})$*. In fact one has even* $\eta_W^{(2N)}(0) < 0$*, so that* $|\eta_W| < 1$ *locally around 0. You can check that the latter statement is true in both the plots of* η_W *of Figure 2, even when* $\|\eta_W\|_\infty > 1$*.*

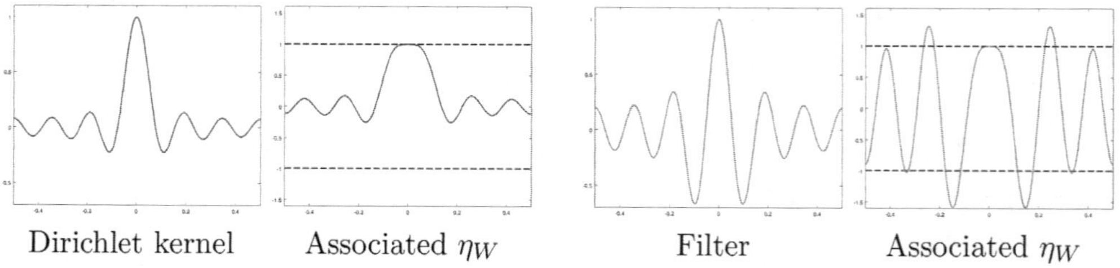

Figure 2. η_W for the ideal low-pass filter (left) and a filter with high extreme Fourier coefficients (right).

2.3. Main Result

The results of the last section enable us to state the following theorem which is our main contribution.

Theorem 1. *If η_W is $(2N-1)$-non-degenerate, then there exist positive constants $t_1, C, M > 0$ (which only depends on the filter and the initial measure) such that:*

$$\forall t \in (0, t_1),\ \forall (\lambda, w) \in B(0, Ct^{2N-1}) \text{ with } \|\frac{w}{\lambda}\|_2 \leq C, \tag{1}$$

there exists a unique measure $m_{a,tz}$ solution of the BLASSO composed of exactly N spikes. Moreover,

$$\|(a,z) - (a_0, z_0)\| \leq M \left(\frac{|\lambda|}{t^{2N-1}} + \frac{\|w\|_2}{t^{2N-1}} \right).$$

The theorem provides an upper-bound on the error of the recovered measure $m_{a,tz}$ with respect to the initial measure m_{a_0,tz_0}. In particular when w/t^{2N-1} and λ/t^{2N-1} tend to zero, $m_{a,tz}$ converges towards m_{a_0,tz_0}.

Figure 3 shows the domain, given in Equation (1), in which there is exact support recovery of the initial measure. Note that when the number of spikes increases, the domain gets smaller i.e. the recovery of the support is harder to perform. This is reflected in Figure 1 where η_W gets flatter around 0 when N increases.

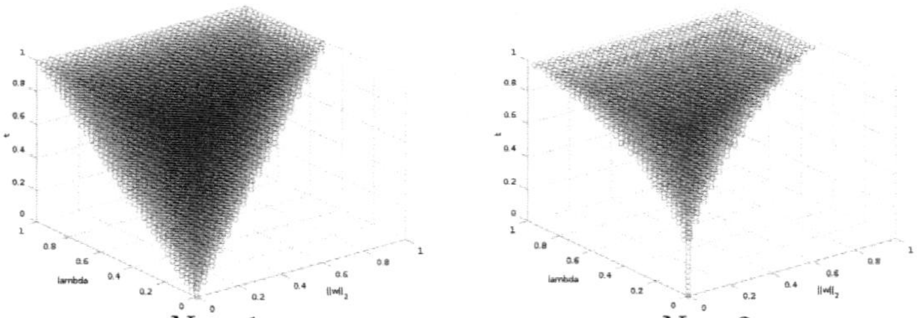

$N = 1$ $N = 2$

Figure 3. Domain of recovery of the support for 1 and 2 spikes.

Acknowledgment

This work has been partly supported by the European Research Council (ERC project SIGMA-Vision).

[1] K. Bredies and H.K. Pikkarainen. Inverse problems in spaces of measures. *ESAIM: Control, Optimisation and Calculus of Variations*, 19:190–218, 2013.

[2] E.J. Candès and C. Fernandez-Granda. Super-resolution from noisy data. *Journal of Fourier Analysis and Applications*, 19(6):1229–1254, 2013.

[3] E.J. Candès and C. Fernandez-Granda. Towards a mathematical theory of super-resolution. *Communications on Pure and Applied Mathematics*, 67(6):906–956, 2014.

[4] Y. de Castro and F. Gamboa. Exact reconstruction using beurling minimal extrapolation. *Journal of Mathematical Analysis and Applications*, 395(1):336 – 354, 2012.

[5] V. Duval and G. Peyré. Exact support recovery for sparse spikes deconvolution. *To appear in Foundation and Computational Mathematics*, 2013.

[6] C. Fernandez-Granda. Support detection in super-resolution. *Proc. Proceedings of the 10th International Conference on Sampling Theory and Applications*, pages 145–148, 2013.

[7] Y. de Castro J.-M. Azaïs and F. Gamboa. Spike detection from inaccurate samplings. *Applied and Computational Harmonic Analysis*, 2014.

[8] V. I. Morgenshtern and E. J. Candès. Stable super-resolution of positive sources: the continuous setup. *in final preparation*, 2014.

[9] G. Peyrè Q. Denoyelle, V. Duval. Asymptotic of sparse support recovery for positive measures. *in preparation*, 2015.

[10] The recoverability limit for superresolution via sparsity. L. demanet and n. nguyen. *preprint*, 2014.

[11] D.L. Donoho S.S. Chen and M.A. Saunders. Atomic decomposition by basis pursuit. *SIAM Journal on Scientific Computing*, 20(1):33–61, 1998.

Journal of Physics: Conference Series **657** (2015) 012014 doi:10.1088/1742-6596/657/1/012014

Reconstruction of 3-D Microwave Images based on a Block-BiCGStab Algorithm

Corentin Friedrich[1,2], Sébastien Bourguignon[1], Jérôme Idier[1] and Yves Goussard[2]

[1] LUNAM Université, École Centrale de Nantes / IRCCyN UMR CNRS 6597, 1 rue de la Noë, 44321 Nantes, France

[2] École Polytechnique de Montréal, Dept. of Electrical Engineering, C.P. 6079, H3C 3A7, Montréal QC, Canada

E-mail: [1]`surname.name@irccyn.ec-nantes.fr`, [2]`yves.goussard@polymtl.ca`

Abstract. In 3-D microwave imaging, gradient-based optimization algorithms usually make use of the so-called stabilized version of the biconjugate gradient iterative method (BiCGStab) in order to solve multiple linear systems. We propose to use a block version of BiCGStab to jointly solve the mutiple right-hand side linear systems. Illuminations are partitioned in subgroups, which makes the method more efficient. The reconstruction process is studied on realistic simulated data and illustrates the efficiency of the method compared to BiCGStab.

1. Introduction

The development of microwave imaging algorithms has received a lot of interest in the last few years with applications such as biomedical imaging and geoscience [1, 2]. Solving the inverse scattering problem under realistic condtions presents several difficulties, a critical one being the reduction of the computational cost associated with three-dimensional problems. Nonlinear inversion methods usually rely on iterative algorithms to reconstruct the dielectric properties of the unknown object. A regularized data misfit cost function is minimized using a gradient-based optimization procedure, such as the Distorted Born iterative method (DBIM) [3] or the Gauss-Newton method [2, 4]. At each iteration, the computation of a descent direction and of a step size requires the evaluation of the objective function and of its gradient. Such computations rely on the resolution of linear systems equal to the number of illuminations. In practice, exact inversion of such systems becomes computationally prohibitive for large 3-D problems and approximate solutions are obtained with iterative methods, such as the conjugate gradient method [5], the quasi minimal residual (QMR) method [6] or the stabilized version of the biconjugate gradient stabilized (BiCGStab) method [1, 7].

Different methods have been proposed to solve efficiently such multiple systems, such as the marching-on-anything technique [4] or a massive parallelization scheme [2]. Since the linear systems share the same system matrix, block versions of iterative algorithms are appropriate, which simultaneously solve linear systems with multiple right-hand sides. In [8], a block-QMR algorithm was proposed for microwave imaging. In [9], a block version of BiCGStab was developed for random system inversion, which was shown to outperform the Block-QMR algorithm. In [10], we proposed a Partial-Block BiCGStab procedure to solve multiple forward

Journal of Physics: Conference Series **657** (2015) 012014 doi:10.1088/1742-6596/657/1/012014

problems in 3-D microwave imaging, which partitions the sources in several subgroups and uses Block-BiCGStab on each group. In this paper, the Partial-Block BiCGStab algorithm is applied for the computation of the objective function and its gradient at each iteration of the inversion process.

In Section 2, the 3-D formulation is presented. The inverse problem is introduced in Section 3 where the Block-BiCGStab algorithm and the Partial-Block version are detailed. Finally, the computational costs of the different methods are compared on simulated data in Section 4.

2. Formulation of the 3-D Problem

An unknown inhomogeneous object of complex permittivity $\epsilon(\boldsymbol{r})$ is included in a volume V, where $\boldsymbol{r}$ denotes the spatial position in V. We assume a homogeneous background medium with complex permittivity ϵ_b and magnetic permeability μ_0, that is the vacuum permeability. We suppose that all objects and media are nonmagnetic. Successive known harmonic incident electric fields $\boldsymbol{E}_i^{\text{inc}}(\boldsymbol{r})$, $i = 1, \ldots, N_S$ illuminate the object with angular frequency ω. The scattered fields $\boldsymbol{E}_i^{\text{scat}}$ are measured at locations $\boldsymbol{r}_j^R$, $j = 1, \ldots, N_R$. In the following, the time dependence term $e^{-j\omega t}$ is omitted. The observation equation links the data with the total electric field $\boldsymbol{E}_i$ in the domain V. It reads [11]:

$$\boldsymbol{E}_i^{\text{scat}}(\boldsymbol{r}_j^R) = \left(k_b^2 + \nabla\nabla\cdot\right) \int_V g(\boldsymbol{r}_j^R, \boldsymbol{r}')\chi(\boldsymbol{r}')\boldsymbol{E}_i(\boldsymbol{r}')d\boldsymbol{r}', \qquad \forall j = 1, \ldots, N_R \tag{1}$$

where k_b and χ respectively denote the wavenumber $k_b = \omega^2\mu_0\epsilon_b$ and the contrast function $\chi(\boldsymbol{r}) = (\epsilon(\boldsymbol{r}) - \epsilon_b)/\epsilon_b$. The $\nabla\nabla\cdot$ term represents the gradient of the divergence. The homogeneous Green's function $g(\boldsymbol{r}, \boldsymbol{r}')$ is defined as $g(\boldsymbol{r}, \boldsymbol{r}') = \exp\left(jk_b\|\boldsymbol{r} - \boldsymbol{r}'\|\right)/4\pi\|\boldsymbol{r} - \boldsymbol{r}'\|$. Note that the integral in (1) is a convolution with the kernel $g(\boldsymbol{r}, \boldsymbol{r}')$. The total field $\boldsymbol{E}_i(\boldsymbol{r})$ in the domain V is ruled by the electric field integral equation (EFIE) or domain equation:

$$\boldsymbol{E}_i(\boldsymbol{r}) = \boldsymbol{E}_i^{\text{inc}}(\boldsymbol{r}) + \left(k_b^2 + \nabla\nabla\cdot\right) \int_V g(\boldsymbol{r}, \boldsymbol{r}')\chi(\boldsymbol{r}')\boldsymbol{E}_i(\boldsymbol{r}')d\boldsymbol{r}', \qquad \forall \boldsymbol{r} \in V. \tag{2}$$

In order to obtain a discrete model, we assume that the volume V is a cuboid divided into N cubic voxels of dimensions N_x, N_y and N_z. Let $\boldsymbol{e}_i^{\text{inc}}$ denote the vector of length $3N$ containing the components of the incident field discretized on each voxel n with center $\boldsymbol{r}_n$, $n = 1, \ldots, N$. Similarly, let $\boldsymbol{e}_i$ denote the discretized total electric field. The contrast function $\chi(\boldsymbol{r})$ is assumed to be constant on each voxel. The vector $\boldsymbol{x}$ of length N contains the unknown contrast on all voxels. The domain equation (2) is discretized using the method of moments [12], which yields:

$$\boldsymbol{e}_i = \boldsymbol{e}_i^{\text{inc}} + \mathbf{G}\mathbf{X}\boldsymbol{e}_i, \tag{3}$$

where $\mathbf{X}$ is a $3N \times 3N$ diagonal matrix whose diagonal contains three replicas of the vector $\boldsymbol{x}$ and $\mathbf{G}$ is a $3N \times 3N$ convolution matrix corresponding to the discretization of $(k_b^2 + \nabla\nabla\cdot)g(\boldsymbol{r}, \boldsymbol{r}')$. The discretization procedure is detailed in [13]. Eq. (3) can be written as a linear system:

$$\boldsymbol{e}_i^{\text{inc}} = \mathbf{L}_{\boldsymbol{x}}\,\boldsymbol{e}_i, \text{ with } \mathbf{L}_{\boldsymbol{x}} = \mathbf{I}_{3N} - \mathbf{G}\mathbf{X} \tag{4}$$

where $\mathbf{I}_{3N}$ is the identity matrix of size $3N$. Determining the total field $\boldsymbol{e}_i$ in (4) corresponds to solving the forward problem.

The observation equation (1) is discretized similarly. The three components of the scattered fields $\boldsymbol{E}_i^{\text{scat}}$ measured at the N_R receivers are stored in the vector $\boldsymbol{y}_i$ of size $3N_R$. The discretized observation equation reads

$$\boldsymbol{y}_i = \mathbf{G}_o\mathbf{X}\,\boldsymbol{e}_i + \boldsymbol{b}. \tag{5}$$

The $N_R \times N$ matrix $\mathbf{G}_o$ is constructed similarly to $\mathbf{G}$. The vector $\boldsymbol{b}$ is a perturbation term accounting for noise and model errors. Combining the domain equation (4) and the observation equation (5), the measurements $\boldsymbol{y}_i$ can be expressed as functions of the contrast $\boldsymbol{x}$ according to:

$$\boldsymbol{y}_i = \mathbf{G}_o \mathbf{X} \mathbf{L}_{\boldsymbol{x}}^{-1} \boldsymbol{e}_i^{\text{inc}} + \boldsymbol{b}. \tag{6}$$

3. Three-dimensional reconstruction

3.1. Optimization of a regularized objective function

Solving the inverse problem consists in determining the contrast function of an unknown object embedded in the volume V from measurements of the scattered field for N_S incident fields. Using the discretized model (6), the problem amounts to estimating the discretized contrast $\boldsymbol{x}$ from data $\boldsymbol{y}_1, \ldots, \boldsymbol{y}_{N_S}$. This problem is ill-posed, therefore we define the solution as the minimizer of a regularized cost function:

$$\widehat{\boldsymbol{x}} = \arg\min_{\boldsymbol{x}} \mathcal{F}(\boldsymbol{x}), \text{ with } \mathcal{F}(\boldsymbol{x}) = \mathcal{F}^D(\boldsymbol{x}) + \lambda \mathcal{R}(\boldsymbol{x}) \tag{7}$$

where $\mathcal{F}^D(\boldsymbol{x}) = \sum_{i=1}^{N_S} \left|\left|\boldsymbol{y}_i - \mathbf{G}_o \mathbf{X} \mathbf{L}_{\boldsymbol{x}}^{-1} \boldsymbol{e}_i^{\text{inc}}\right|\right|^2$ is the quadratic data misfit and $\mathcal{R}(\boldsymbol{x})$ is a regularization term operating on the differences between adjacent voxels:

$$\mathcal{R}(\boldsymbol{x}) = \sum_{k,\ell,m} \varphi(\mathbb{X}_{k,\ell,m} - \mathbb{X}_{k-1,\ell,m}) + \varphi(\mathbb{X}_{k,\ell,m} - \mathbb{X}_{k,\ell-1,m}) + \varphi(\mathbb{X}_{k,\ell,m} - \mathbb{X}_{k,\ell,m-1}) \tag{8}$$

where $\{\mathbb{X}_{k,\ell,m}\}_{k,\ell,m}$ represents the 3-D arrangement of $\boldsymbol{x}$. We use the "$\ell_2\ell_1$" function $\varphi(u) = \sqrt{\delta^2 + |u|^2}$, which tends to preserve edges in the reconstructed 3D-image. Moreover, for any $\delta \neq 0$, the cost function is continuously differentiable on $\mathbb{R}^N$ and gradient-based optimization methods can be used to solve (7). The parameter λ controls the trade-off between the data misfit and the regularization function. In this paper, parameters λ and δ are set empirically. We consider the local minimization of criterion (7) with the L-BFGS algorithm [14]. Each iteration of such algorithm involves one computation of the gradient and possibly several computations of the objective function for the line search step. The gradient reads:

$$\nabla_{\boldsymbol{x}} \mathcal{F} = -\sum_{i=1}^{N_S} \operatorname{diag}\left\{\mathbf{L}_{\boldsymbol{x}}^{-1} \boldsymbol{e}_i^{\text{inc}}\right\}^\dagger \left(\overline{\mathbf{L}_{\boldsymbol{x}}}\right)^{-1} \left[\mathbf{G}_o^\dagger \left(\boldsymbol{y}_i - \mathbf{G}_o \mathbf{X} \mathbf{L}_{\boldsymbol{x}}^{-1} \boldsymbol{e}_i^{\text{inc}}\right)\right] + \nabla_{\boldsymbol{x}} \mathcal{R}, \tag{9}$$

where the symbol † denotes the Hermitian adjoint, the overline denotes the complex conjugation and diag$\{\boldsymbol{u}\}$ stands for the diagonal matrix with diagonal $\boldsymbol{u}$.

3.2. Block algorithms for linear system resolutions

The evaluation of the cost function $\mathcal{F}$ requires the computation of N_S forward problems $\mathbf{L}_{\boldsymbol{x}}^{-1} \boldsymbol{e}_i^{\text{inc}}$, $i = 1, \ldots, N_S$. The computation of the gradient also requires the resolution of N_S systems with matrix $\overline{\mathbf{L}_{\boldsymbol{x}}}$. For realistic 3-D problems, exact inversion of such linear systems is computationally prohibitive: $\mathbf{L}_{\boldsymbol{x}}$ is a full matrix with $3N \times 3N$ elements. Consequently, the systems are generally solved by iterative methods, such as the BiCGStab algorithm [1, 2], which must be called N_S times for the computation of $\mathcal{F}$ and N_S times for the resolution of systems with normal matrix $\overline{\mathbf{L}_{\boldsymbol{x}}}$ in (9). The most computationally demanding steps in BiCGStab are the matrix-vector multiplications with the matrix $\mathbf{L}_{\boldsymbol{x}}$, which essentially correspond to convolution operations, that can be efficiently implemented by Fast Fourier Transform algorithms.

In this paper, we propose a method to solve all the linear systems jointly. Indeed, the operator $\mathbf{L}_{\boldsymbol{x}}$ does not depend on the illuminations and the linear systems can be aggregated. The forward problems involved in the computation of both the cost function and its gradient read:

$$\mathbf{E} = \mathbf{L}_{\boldsymbol{x}}^{-1} \mathbf{E}^{\text{inc}} \tag{10}$$

with $\mathbf{E}^{\mathrm{inc}} = [\boldsymbol{e}_1^{\mathrm{inc}}, \ldots, \boldsymbol{e}_{N_S}^{\mathrm{inc}}]$ and $\mathbf{E} = [\boldsymbol{e}_1, \ldots, \boldsymbol{e}_{N_S}]$. The computation of the gradient also requires a second set of system resolutions with matrix $\overline{\mathbf{L}_{\boldsymbol{x}}}$, that can also be aggregated as:

$$\overline{\mathbf{L}_{\boldsymbol{x}}}^{-1}\left[\mathbf{G}_o^{(3)\dagger}\left(\mathbf{Y} - \mathbf{G}_o^{(3)}\mathbf{X}^{(3)}\mathbf{L}_{\boldsymbol{x}}^{-1}\mathbf{E}^{\mathrm{inc}}\right)\right] \tag{11}$$

with $\mathbf{Y} = [\boldsymbol{y}_1, \ldots, \boldsymbol{y}_{N_R}]$. We propose to use the Block-BiCGStab algorithm, which is a generalization of the BiCGstab algorithm to multiple right-hand side systems [9]. It is described in Algorithm 1 for the joint resolution of the forward problems in (10). The notation $\langle \mathbf{T}, \mathbf{S} \rangle_F$ denotes the Frobenius product between matrices $\mathbf{T}$ and $\mathbf{S}$, that is, the trace of matrix $\mathbf{T}^\dagger\mathbf{S}$. If $N_S = 1$, Block-BiCGStab reduces to BiCGStab. In the following, the implementation of the computation of the total fields in (10) is detailed. The resolution of (11) is performed similarly.

Algorithm 1 Block-BiCGStab algorithm

1: For an initial guess $\mathbf{E}^{(0)}$, $\mathbf{R}^{(0)} = \mathbf{E}^{\mathrm{inc}} - \mathbf{L}_{\boldsymbol{x}}\mathbf{E}^{(0)}$, $\mathbf{P}^{(0)} = \mathbf{R}^{(0)}$
2: Choose an arbitrary $3N \times N_S$ matrix $\widetilde{\mathbf{R}}_0$
3: $k = 0$
4: **repeat**
5: $\quad \mathbf{V} = \mathbf{L}_{\boldsymbol{x}}\mathbf{P}^{(k)}$
6: $\quad$ solve $(\widetilde{\mathbf{R}}_0^\dagger\mathbf{V})\mathbf{A} = \widetilde{\mathbf{R}}_0^\dagger\mathbf{R}^{(k)}$
7: $\quad \mathbf{S} = \mathbf{R}^{(k)} - \mathbf{V}\mathbf{A}$
8: $\quad \mathbf{T} = \mathbf{L}_{\boldsymbol{x}}\mathbf{S}$
9: $\quad \omega = \langle \mathbf{T}, \mathbf{S} \rangle_F / \langle \mathbf{T}, \mathbf{T} \rangle_F$
10: $\quad \mathbf{E}^{(k+1)} = \mathbf{E}^{(k)} + \mathbf{P}^{(k)}\mathbf{A} + \omega\mathbf{S}$
11: $\quad \mathbf{R}^{(k+1)} = \mathbf{S} - \omega\mathbf{T}$
12: $\quad$ solve $(\widetilde{\mathbf{R}}_0^\dagger\mathbf{V})\mathbf{B} = -\widetilde{\mathbf{R}}_0^\dagger\mathbf{T}$
13: $\quad \mathbf{P}^{(k+1)} = \mathbf{R}^{(k+1)} + (\mathbf{P}^{(k)} - \omega\mathbf{V})\mathbf{B}$
14: $\quad k \leftarrow k + 1$
15: **until** $\forall i, \|\boldsymbol{r}_i^{(k)}\| / \|\boldsymbol{e}_i^{\mathrm{inc}}\| <$ tolerance, where $\boldsymbol{r}_i^{(k)}$ is the i-th column of $\mathbf{R}^{(k)}$

Algorithm 2 Partial-Block BiCGStab algorithm

1: Build subsets $\mathbf{E}^{\mathrm{inc}(p)}$ for $p = 1 \ldots P$ by partitioning the N_S sources.
2: **for** $p = 1 \ldots P$ **do**
3: $\quad$ Run the Block-BiCGStab algorithm on $\mathbf{E}^{\mathrm{inc}(p)}$ and $\mathbf{E}^{(p)}$ instead of $\mathbf{E}^{\mathrm{inc}}$ and $\mathbf{E}$.
4: **end for**

The Block-BiCGStab method is an iterative algorithm, where each iteration requires the resolution of two linear systems of size N_S with matrix $\widetilde{\mathbf{R}}_0^\dagger\mathbf{V}$ at steps 6 and 12 of Algorithm 1. These systems are relatively small, therefore the corresponding matrices can be computed and system resolutions can be performed exactly. However, during the iterative process, the matrix $\widetilde{\mathbf{R}}_0^\dagger\mathbf{V}$ may become ill-conditioned, which leads to instabilities. This may happen in particular when two sources, say with indices i and j, are very close. In that case, the incident fields $\boldsymbol{e}_i^{\mathrm{inc}}$ and $\boldsymbol{e}_j^{\mathrm{inc}}$ are highly correlated and this property is propagated into the quantities $\mathbf{R}^{(k)}$ and $\mathbf{V}$. To overcome the instabilities, we propose a procedure called Partial-Block BiCGStab. To avoid ill-conditioned matrices, the N_S sources are divided into P subgroups such that all sources in one subgroup p are as far as possible from each other. Then, matrix $\mathbf{E}^{\mathrm{inc}}$ is split into P submatrices $\mathbf{E}^{\mathrm{inc}(p)}$ in which the columns are weakly correlated. Similarly, matrix $\mathbf{E}$ is split into submatrices $\mathbf{E}^{(p)}$. Then, the Block-BiCGStab is applied independently on the P subproblems. This procedure is summarized in Algorithm 2. For $P = 1$, Partial-Block BiCGStab reduces to Block-BiCGStab. For $P = N_S$, it reduces to the BiCGStab algorithm. The choice of P is important, more information on its influence are given in [10]. The resolution of the multiple forward problems is faster for Partial-Block BiCGStab than for BiCGStab and is even better for high contrasted objects [10]. Here, the Partial-Block BiCGStab procedure is used to compute the cost function and the gradient as well in order to speed up the reconstruction algorithm using L-BFGS.

Journal of Physics: Conference Series **657** (2015) 012014 doi:10.1088/1742-6596/657/1/012014

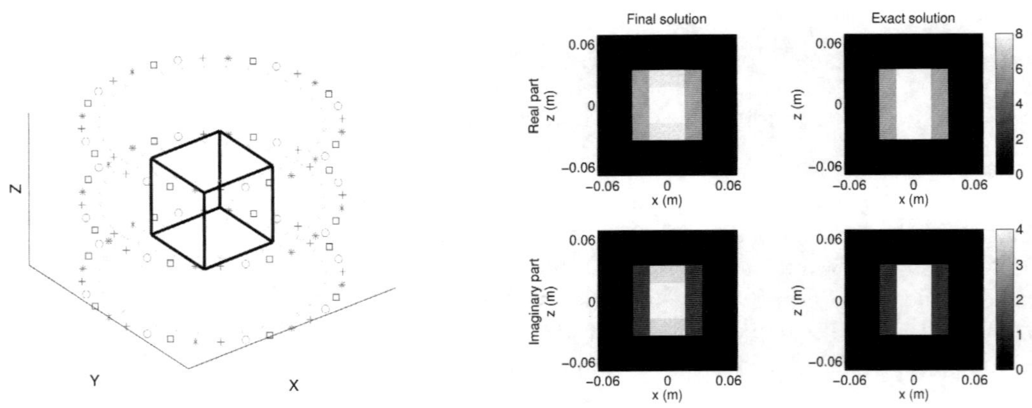

Figure 1. Left: configuration setup with 96 sources and 96 receivers (cyan x symbols) distributed on three coaxial circles. The black cube represents the volume of interest V. The four different symbols ($+$, $\star$, $\circ$, $\square$) represent the four subgroups of sources.
Right: cross-section of the object at $y = 0$. First colum: real and imaginary parts of the obtained solution. Second column: true object.

4. Simulation Results

In this section, we evaluate the computation time and the convergence of our method compared to BiCGStab and Block-BiCGStab. First, the resolution of the multiple forward problems are studied. Then, results on the reconstruction process are presented. We consider a simulated experiment presented in Fig. 1 (left) that can be found in the literature, with $N_S = 96$ sources at frequency 2.45 GHz embedded in the air of permittivity $\epsilon_b = 1$. The sources are placed on three coaxial circles at heights $z = -\lambda_b$, 0 and λ_b, and radius $1.5\lambda_b$, where λ_b is the wavelength of the incident fields. On each circle, $N_S/3 = 32$ sources are equally distributed. A lossy cube of size $0.5\lambda_b$ with contrast $\chi_1 = 6 + 1.2j$ is embedded in a cubic volume V of size λ_b. Inside the lossy cube, a cuboid is inserted with square cross-section of $0.25\lambda_b$, height $0.5\lambda_b$ and contrast $\chi_2 = 8 + 4j$. The volume V is discretized into $8 \times 8 \times 8$ voxels. The 96 receivers are placed on three coaxial circles at the same heights as the sources and with radius $1.3\lambda_b$.

Synthetic data are simulated on a grid of $16 \times 16 \times 16$ voxels and zero-mean white Gaussian noise is added with signal-to-noise ratio equal to 20 dB. The reconstruction procedure is implemented with Matlab on a personal computer with 8 Go RAM and eight cores clocked at 3.40 GHz (four hyperthreaded cores). The reconstruction process is parallelized by simply executing loops with Matlab `parfor` instruction. For BiCGStab, the $2N_S$ system resolutions are sent in parallel to the different processors. The Block-BiCGStab algorithm is not easily parallelizable and therefore is not parallelized. For the Partial-Block BiCGStab algorithm, the resolutions of the different subproblems are also sent in parallel to the different processors. For Partial-Block BiCGStab, we divide the set of sources into four subgroups (see Fig. 1 left). For Block-BiCGStab and Partial-Block BiCGStab, $\widetilde{\mathbf{R}}_0$ is set to $\mathbf{L}_{\boldsymbol{x}}\mathbf{R}^{(0)}$.

Fig. 2 plots the evolution of the relative residual errors for the resolution of the 96 forward problems using the three algorithms: BiCGStab, Block-BiCGStab and Partial-Block BiCGStab. The tolerance is set to 10^{-6}, which is a strict value. Two initializations are tested: $\mathbf{E}^{(0)} = \mathbf{E}^{\text{inc}}$, and $\mathbf{E}^{(0)}$ close to the exact solution of (10) (which can be computed directly for such reasonable size problem). We note that regardless of the initialization, BiCGStab and Partial-Block BiCGStab converge in nearly the same number of iterations, but Partial-Block BiCGStab is the most efficient. On the contrary, the Block-BiCGStab algorithm converges much faster if it is initialized close to the solution. The computation times for the three resolutions with

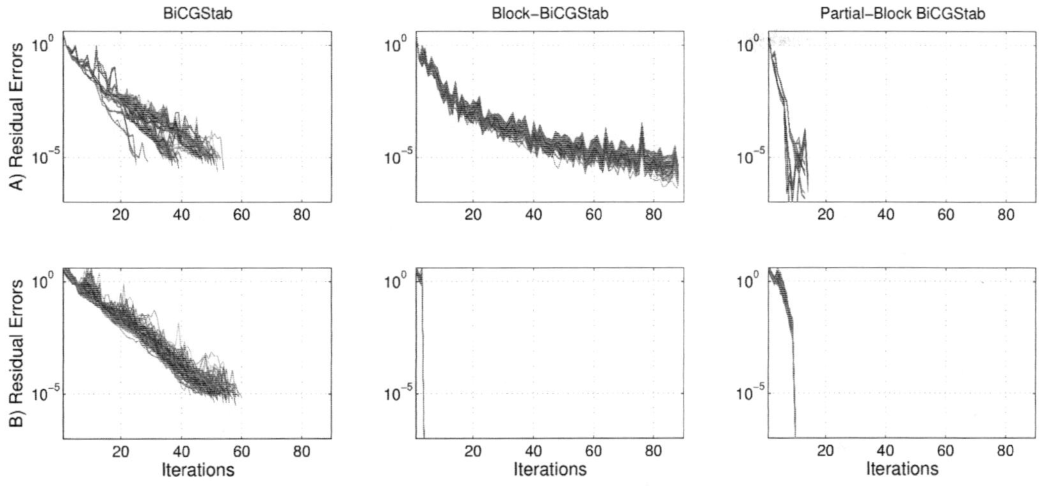

Figure 2. Residual errors for the 96 forward problems as a function of iterations for BiCGStab, Block-BiCGStab and Partial-Block BiCGStab. First row: algorithms are initialized with the incident fields $\mathbf{E}^{(0)} = \mathbf{E}^{\text{inc}}$. Second row: algorithms are initialized close to the solution.

$\mathbf{E}^{(0)} = \mathbf{E}^{\text{inc}}$ are respectively 17 s, 38 s and 6.5 s. With parallelization, the cost of BiCGStab decreases to 4.3 s and that of Partial-Block BiCGStab to 2.2 s.

In the reconstruction process, the three algorithms were implemented to solve (10) and (11) involved in the cost function and its gradient. The hyperparameter λ was set to 10^{-5} and δ to 10^{-2}. At iteration k of the L-BFGS algorithm, the resolutions of (10) and (11) are initialized with the solutions obtained at iteration $k-1$. Implementations based on the BiCGStab, Block-BiCGStab and Partial-Block BiCGStab algorithms converge respectively in 855 s, 1004 s and 573 s. The tolerance set to 10^{-6} on the residual errors for the three algorithms leads to good accuracy in the resolutions of the linear systems.

A perspective is to solve inverse problems of larger size. We will also study the influence of the tolerance: a weak tolerance generates strong approximations in the system resolutions and could introduce instabilities in the convergence of the reconstruction algorithm.

References

[1] Zhang Z Q, Liu Q H, Xiao C, Ward E, Ybarra G and Joines W T 2003 IEEE Trans. Biomed. Eng. **50** 1180–1189
[2] Abubakar A, Habashy T M, Pan G and Li M K 2012 IEEE Trans. Ant. Propag. **60** 2431–2441 ISSN 0018-926X
[3] Winters D W, Van Veen B D and Hagness S C 2010 IEEE Trans. Ant. Propag. **58** 145–154 ISSN 0018-926X
[4] De Zaeytijd J, Franchois A, Eyraud C and Geffrin J M 2007 IEEE Trans. Ant. Propag. **55** 3279–3292
[5] Zwamborn P and van den Berg P M 1992 IEEE Trans. Microwave Theory Tech. **40** 1757–1766
[6] Wang C F and Jin J M 1998 IEEE Trans. Microwave Theory Tech. **46** 553–558 ISSN 0018-9480
[7] Van der Vorst H A 1992 SIAM J. Stat. Sci. Comp. **13** 631–644
[8] Fang Q, Meaney P M, Geimer S D, Streltsov A V and Paulsen K D 2004 IEEE Trans. Medical Imaging **23** 475–484
[9] El Guennouni A, Jbilou K and Sadok H 2003 Electron. Trans. Numer. Anal. **16** 2
[10] Friedrich C, Idier J, Bourguignon S and Goussard Y 2015 Accep. for pub. Proc. 9th European Conf. Antennas Propag.
[11] Pastorino M 2010 Microwave Imaging (John Wiley & Sons)
[12] Harrington R F and Harrington J L 1996 Field Computation by Moment Methods (Oxford University Press)
[13] Abubakar A and van den Berg P M 2004 J.Comput. Phys. **195** 236–262 ISSN 0021-9991
[14] Bertsekas D P 1999 Nonlinear Programming (Belmont, MA: Athena Scientific)

Surface crack reconstruction from eddy current images using a direct semi-analytic model

Eric Vourc'h[1], Thierry Bore[1], Caifang Cai[2], Romain Soulat[3]
[1]SATIE (ENS Cachan/CNRS/Paris-Saclay University), 61 Avenue du Président Wilson, 94230 Cachan, France
[2]L2S, CentraleSupelec-CNRS-Univ. Paris-Sud, 3 Rue Joliot-Curie, 91192 Gif sur Yvette, France
[3]LSV (ENS Cachan/CNRS/Paris-Saclay University), 61 Avenue du Président Wilson, 94230 Cachan, France
E-mail: eric.vourch@satie.ens-cachan.fr, bore.thierry@gmail.com, caifang.cai@lss.supelec.fr, romain.soulat@thalesgroup.com

Abstract. A method relying on a direct semi-analytic model is proposed for reconstructing cracks from eddy current images. We consider systems featuring a uniform excitation flow which can be modelled by means of fictitious current sources distributed in the crack volume. Thanks to the relative simplicity of the model a reconstruction method based the comparison of EC images is considered. The sensitivity of different images comparison criteria is studied for 2D surface cracks, leading to a reconstruction method based on a genetic algorithm. Numerical experiments are carried out to examine the performances of the algorithm in terms of convergence and accuracy.

1. Introduction

Non destructive evaluation (NDE) contributes to the security of industrial facilities as a means of preventing accidents by early detecting and characterizing defects. With ultrasounds [1], eddy current (EC) methods [2] are among the most widespread NDE techniques as they are particularly useful for evaluating corrosion and cracks in electrically conductive structures. However, the accurate reconstruction of the shape of cracks and defects by means of EC data is still an open scientific stake because the inversion of EC data is an ill-posed problem [3-5].

To cope with this issue several approaches are therefore investigated, among which total variation regularisation methods [6] or the use a fast interpolation models (such as meta-models obtained from databases) and the estimation of a restricted amount of parameters characterising the shape of the defects [7]. Since a main difficulty of the inversion relies on the establishment of a confidence criterion on the results of the algorithm, statistical approaches (such as Monte Carlo [8]) also are considered which require a high number of resolutions of the direct problem. Non-iterative algorithms also are envisaged based on punctual tests of the inspected structure with near real-time NDE applications in view [9].

An NDE technique usually involves both a sensing device (composed of the excitation and sensing elements) and computations. With a view to optimize the coupling of the excitation signal in the inspected structure and the characteristics of the sensed one, the EC device shall be designed taking the geometry and characteristics of the inspected structure into account. It is also desirable to adapt the inversion method and the experimental setup to each other; which understates the adaptation of the

Journal of Physics: Conference Series **657** (2015) 012015 doi:10.1088/1742-6596/657/1/012015

modelling of the EC problem when the inversion method draws on a model. Such an approach is followed in this paper.

We consider EC sensing systems relying on harmonic excitation and designed to induce a uniform EC flow within the inspected structure (as opposed to circular excitation configurations) [10]. Such systems are sensitive to the presence of cracks and generate a significant magnetic field perpendicular to the surface of the inspected structure. Furthermore, it has been demonstrated that the interactions between a uniform EC flow and a crack could suitably and rather simply be modelled by means of fictitious current sources placed within the volume of the crack [11]. Moreover, the implementation of this model by means of the distributed point source method (DPSM), which is a generic semi-analytic approach, can provide with fast computation [12].

Armed with this direct model, we propose a crack reconstruction method based the comparison of EC images (i.e. cartographies of the magnetic field component perpendicular to the surface of the structure), focusing on the reconstruction of 2D surface cracks.

The paper is organized as follows. Section 2 reports on the description of the direct model and its implementation using DPSM. In section 3 the sensitivity of different image comparison criteria to the dimensions of surface cracks is studied, leading to a reconstruction method based on a genetic algorithm (GA) [14]. In section 4, a GA is applied to the reconstruction of 2D surface cracks.

2. The direct model

The modelling of the interactions between a uniform EC flow and a crack allows for certain simplifications: assuming a superposition hypothesis, the EC surrounding the crack induce a magnetic field distribution similar to the one virtual EC flowing within the crack volume would induce provided they are equal in amplitude and in phase opposition to those which would be present in the same volume in a crack-free material [11]. This model implies that the crack is of small dimensions and represents a small perturbation of the EC flow [13], which means that the amount of EC deviated by the crack is negligible. However approached, this model can prove to be sufficient [11]. Moreover, with the inversion of EC data in view, even more so for iterative inversion methods, its use may be worth considering as it should require a limited computation time.

The model lends itself to an implementation by means of the DPSM, a semi-analytic approach for which the active sources present in the workspace are modelled by a discrete ensemble of elementary radiating sources (according to Green functions). In addition to the active sources, the DPSM implements virtual interface sources calculated according to the boundary conditions. Only the location of the active and virtual interface sources and those where the physical quantities are wished to be computed (as the superposition of the contributions of the radiating sources) require to be meshed, which makes the DPSM computationally efficient.

We will not go into the details of the DPSM here, a comprehensive description being available in [12]. Our purpose will be limited to the main elements of the implementation of the simplified EC modelling proposed above (Figure 1). In that case, the workspace is divided into two media: air and metal, and the physical quantity of interest (i.e. the distribution of the vertical magnetic field $\boldsymbol{b}_z$ in the air above the surface of the metallic structure) calculated as the sum of the contributions of the virtual interface sources radiating in the air. In practice, the problem is formulated in such a way that $\boldsymbol{b}_z$ is computed as the product of a matrix by the current sources placed in the crack volume.

It is to be noted that for the sake of accuracy of the modelling it is suitable that the interface be meshed respecting a distance between the sources lower than a third of the characteristic distance of the electromagnetic problem, which is typically in the order of the skin depth δ:

$$\delta = 1/\sqrt{\pi f \sigma \mu_0 \mu_r} \qquad (1)$$

where f is the frequency of the EC, μ_0 is the magnetic permeability in a vacuum, μ_r is the relative magnetic permeability and σ is the electric conductivity of the medium.

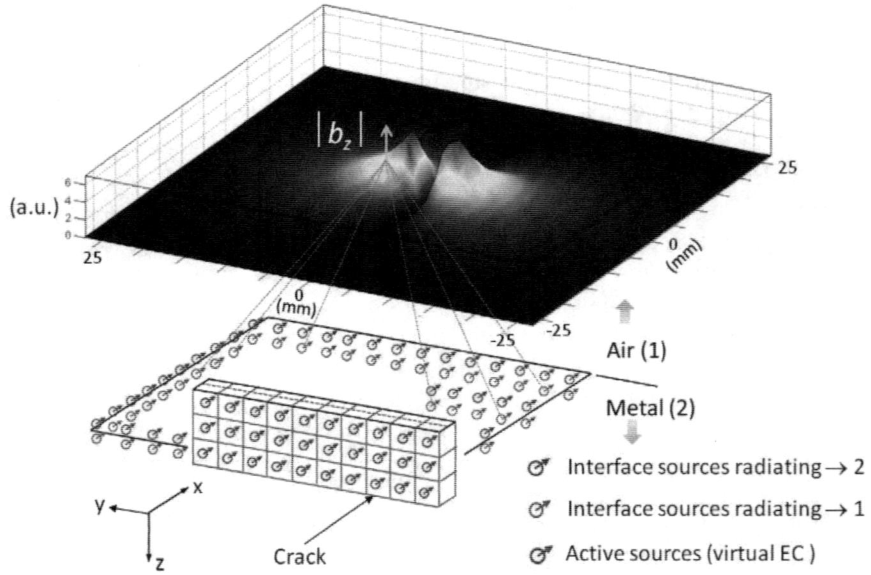

Figure 1. DPSM modelling of the interactions between a uniform EC flow and a 3D crack featuring a parallelepiped shape; cartography of the modulus of the vertical component of the magnetic field above the surface of the metallic part.

3. EC images parametric study

The computational efficiency of the simplified EC direct model proposed above makes it reasonable to envisage iterative crack reconstruction methods relying on the comparison of EC images obtained from the direct model. We focus here on the development of such an approach applied to the reconstruction of surface cracks. Previous to the design of a 2D reconstruction algorithm, we analyze the relevance of different criteria of image comparison.

The EC model is implemented via DPSM with the discretization configuration described in Table 1. The metal is assumed to be an aluminum alloy featuring σ= 16 MS.m^{-1} and μ_r = 1. f is fixed at 400 Hz and the meshing of the interface fulfils the rule stated above, the interface sources being spaced out of $\delta/3$ ($\simeq$ 6 mm). The EC flow along the x axis like in Figure 1.

Table 1. DPSM configuration.

	z position	x dimension		y dimension	
	(mm)	(mm)	(pixels)	(mm)	(pixels)
Active sources plane	-2	20	n = 40	20	m = 40
b_z cartography	0	54	n = 40	54	m = 40

Let us denote $\boldsymbol{b_z} \in \mathbb{C}^M$ the vertical magnetic field at the surface of the metallic structure, where $M = m \times m$ is the image (i.e. cartography) dimension. Let us consider 3 criteria of comparison of images: a correlation error ε_ρ, a normalized mean square error ε_{mse} and a normalized weighted mean square error ε_{wmse} defined as relative errors between a magnetic field $\boldsymbol{b_z}$ and a reference $\boldsymbol{b_{z0}}$:

$$\varepsilon_\rho = \frac{[\boldsymbol{b_z}-E(\boldsymbol{b_z})]'[\boldsymbol{b_z}-E(\boldsymbol{b_{z0}})]}{\sqrt{[\boldsymbol{b_z}-E(\boldsymbol{b_z})]'[\boldsymbol{b_z}-E(\boldsymbol{b_z})]}\cdot\sqrt{[\boldsymbol{b_{z0}}-E(\boldsymbol{b_{z0}})]'[\boldsymbol{b_{z0}}-E(\boldsymbol{b_{z0}})]}} \quad (2)$$

$$\varepsilon_{mse} = \frac{[\boldsymbol{b_z}-\boldsymbol{b_{z0}}]'[\boldsymbol{b_z}-\boldsymbol{b_{z0}}]}{\boldsymbol{b_{z0}}'\boldsymbol{b_{z0}}} \quad (3)$$

$$\varepsilon_{wmse} = \frac{[\boldsymbol{b_z}-\boldsymbol{b_{z0}}]'W[\boldsymbol{b_z}-\boldsymbol{b_{z0}}]}{\boldsymbol{b_{z0}}'W\boldsymbol{b_{z0}}} \tag{4}$$

where $W = diag|\boldsymbol{b_{z0}}|$ is the diagonal matrix supporting $|\boldsymbol{b_{z0}}|$ on the diagonal. In (3) and (4), the normalization is performed with respect to the total energy (or weighted energy) of the reference $\boldsymbol{b_{z0}}$.

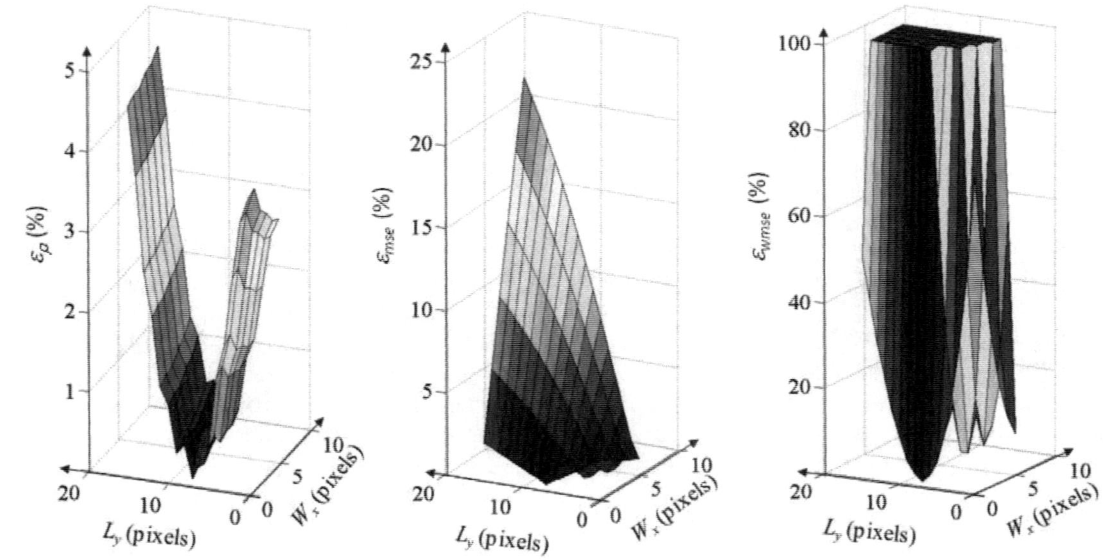

Figure 2. ε_ρ, ε_{mse} and ε_{wmse} for 2D rectangular cracks compared to a 0.5 mm by 4 mm reference crack (1 pixel ⇔ 0.5 mm).

Figure 2 illustrates the sensitivity of ε_ρ, ε_{mse} and ε_{wmse} to the width W_x and length L_y of a rectangular surface crack compared to a 0.5 mm wide and 4 mm long reference. Hence, every error is null at $\{W_x = 0.5$ mm, $L_y = 4$ mm$\}$. It appears that ε_ρ is rather sensitive to L_y and much less to W_x. Indeed, the correlation is reputed sensitive to the variations of the shape of the images, which are sharper along the longitudinal direction of the crack (as shown in Figure 1). Compared to ε_ρ, the variations of ε_{mse} happen to be 5 times higher. Moreover this criterion is sensitive both to W_x and L_y. Regarding ε_{wmse}, a much higher dynamic of variations is observed and the criterion is sensitive both to W_x and L_y. Such a result is consistent with the fact that the higher the signal corresponding to a pixel, the higher its weight (4). This suggests that ε_{wmse} could be the most relevant criterion with implementation in a 2D reconstruction algorithm in view. Furthermore, it is to be noted that the three errors feature multiple minima. It follows that a reconstruction based on a GA seems more suitable than a gradient algorithm since the room left to random evolution of the individuals in a GA enables minimizing the risk of converging towards a local minimum.

4. Surface crack reconstruction

Considering the reconstruction of rectangular surface cracks (and assuming such a feature as an a priori for the reconstruction), we have implemented a GA [14] involving a population of *4* individuals, i.e. rectangular. It is to ne noted that the population size was not optimized for minimizing the number of iterations, but simply chosen as the "magic number" 4. At each iteration of the GA, the fitness of the individuals is evaluated thanks to an objective function based on a comparison between the EC images $\boldsymbol{b_z}$ of the individuals (obtained from the direct model described above) and the image $\boldsymbol{b_{z0}}$ of the crack to be reconstructed. A selection of the individuals is performed according to a fortune wheel and followed by a mutation of the population (by randomly varying according to a given Gaussian law the dimensions of the selected individuals). The initialization of the population consists in randomly choosing the individuals while ensuring a minimum fitness of each individual.

Table 2. GA convergence statistics.

Objective function	Average number of iterations	Standard deviation
ε_ρ	23.13	20.10
ε_{mse}	22.69	17.68
ε_{wmse}	16.83	13.65

Statistics based on 100 drawings of the GA for each objective function. Researched crack / $\{W_x = 0.5$ mm, $L_y = 4$ mm$\}$ with $\boldsymbol{b}_{z0}$ featuring 30 dB SNR.

Convergence statistics of the GA towards the right rectangular defect were performed for objective functions based on each of the 3 image comparison criteria considered in section 3 (Table 2). Whatever the objective function the GA converges towards the right $\{W_x, L_y\}$. Nonetheless, the objective function based on ε_{wmse} requires less iterations than the others to converge, which is consistent with the results of the previous sensitivity study.

To further examine the estimation performances, numerical experiments have been carried out with only 15 iterations per drawing, which is lower than the average number required for the GA to converge. 250 drawings of the GA were made with each objective function and for 5 signal to noise rations (SNR) (Figure 3). Whatever the objective function the resulting average errors are very small whether regarding $\hat{L}_y$ or $\widehat{W}_x$, even though the SNR considered is low. However, the results tend to confirm that the objective function based on ε_ρ provides with more accurate estimation of L_y than W_x, while using ε_{wmse} leads to the overall most accurate estimation of both L_y and W_x.

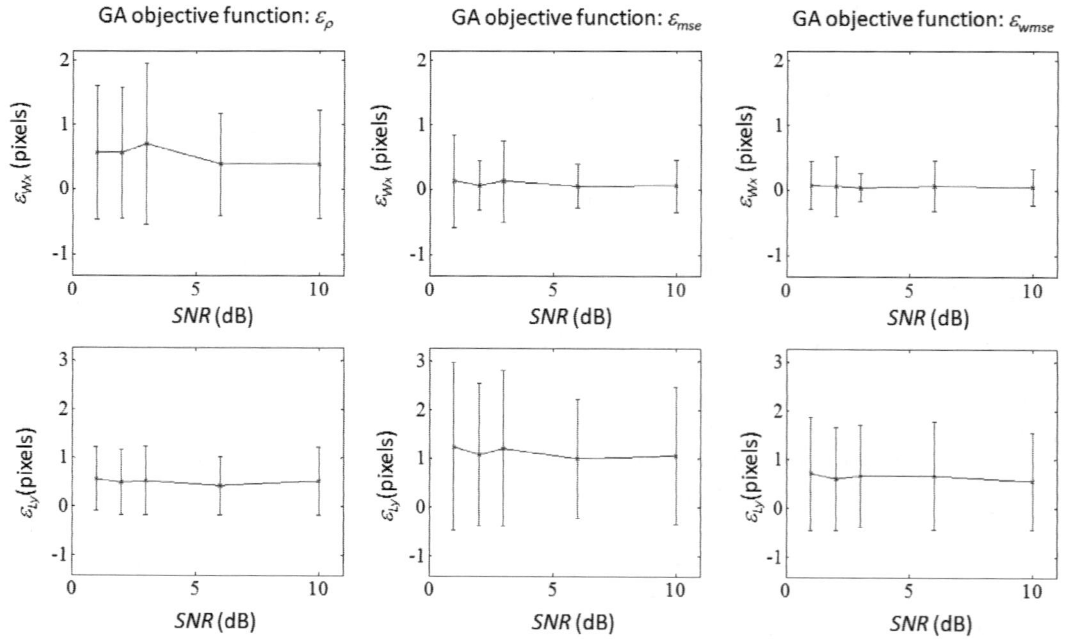

Figure 3. Errors ε_{W_x} and ε_{L_y} (expectancy and standard deviation of $\widehat{W}_x$ and $\hat{L}_y$ for 250 drawings of the GA limited to 15 iterations). $\{W_x = 0.5$ mm, $L_y = 4$ mm$\}\Leftrightarrow\{W_x = 1$ pixel, $L_y = 8$ pixels$\}$.

Another aspect of numerical experiments is the computational efficiency of the reconstruction method. Still in the conditions described in section 3, the computation time of the DPSM transfer matrix (calculated once for all prior to iterating the GA) was in the order of 12 seconds, while an iteration of the GA did last about 3.5 seconds. The algorithm was implemented by means of Matlab™ on a standard PC.

5. Conclusion

In this work, an iterative approach based on the use of a semi-analytic direct model of EC images has been proposed for reconstructing surface cracks. The method which applies to systems featuring homogeneous EC induction has been tested via a GA with simulated data in the presence of noise. Convergence of the GA towards the exact solution was obtained for several fitness functions based on different criteria of comparison of EC images. The relevance of the considered criteria to the estimation of the length and width of cracks also was studied by implementing the GA with a small number of iterations.

Based on these results the reconstruction of 3D cracks (either surface or buried cracks) can be envisaged. Therefore, further works will focus on a multi-frequency approach taking the skin effect into account in order to reconstruct the depth of structures layer by layer. The application of the method to experimental data will also be considered.

6. References

[1] Kubinyi M, Docekal A, Ramos H and Ribeiro A 2010 *Przegląd Elektrotechniczny* **86** 249-254

[2] Bernieri A, Ferrigno L, Laracca M and Molinara M 2008 *Instrumentation and Measurement, IEEE Transactions on* **57**(9) 1958-1968

[3] Colton D and Päivärinta L 1992 *Archive for rational mechanics and analysis* **119**(1), 59-70

[4] Isakov, V. 1993 *Inverse Problems* **9**(6) 579

[5] Yamamoto M 1997 *International Journal of Applied Electromagnetics and Mechanics* **8**(1) 77-98

[6] Ribeiro A L, Ramos H G, Pasadas D and Rocha T 2014, February *40th Annual Review Of Progress In Quantitative Nondestructive Evaluation AIP Publishing* **1581**(1) 1428-1432

[7] Douvenot R, Lambert M and Lesselier D 2011 April Magnetics, *IEEE Transactions on* **47**(4) 746-755

[8] Cai C, Rodet T and Lambert M 2014 *Journal of Physics: Conference Series,* IOP Publishing **542**(1) 012009

[9] Tamburrino A, Calvano F, Ventre S. and Rubinacci G 2012 *NDT & E International* **47** 26-34.

[10] Koyama K, Hoshikawa H and Taniyama N 2000, October *Proceedings of WCNDT*

[11] Thomas V, Joubert P Y, Vourc'h E and Placko D 2010, February *Sensors Applications Symposium (SAS IEEE)* 154-157

[12] Placko D, Bore T and Kundu T 2012 *Ultrasonic and Electromagnetic NDE for Structure and Material Characterization: Engineering and Biomedical Applications CRC Press* 249-294

[13] Sabbagh H and Sabbagh L 1986 *Magnetics, IEEE Transactions on* **22**(4) 282-291

[14] Goldberg D E and Holland J H 1988 *Machine Learning, Kluwer Academic Publishers* **3**(2-3), 95-99.

Institute of Physics
Dirac House, Temple Back
Bristol BS1 6BE UK

ISSN: 1742-6588
ISBN 978-1-5108-1731-9

CURRAN ASSOCIATES INC.
proceedings.com